4000 MORE FACTS

Bardfield Press is an imprint of Miles Kelly Publishing
Bardfield Centre, Great Bardfield, Essex, CM7 4SL

This edition first published in 2004
(originally published in hardback, 2000)
by Miles Kelly Publishing Ltd
Copyright © Miles Kelly Publishing 2000

2 4 6 8 10 9 7 5 3 1

British Library Cataloguing-in-Publication Data
A catalogue record for this book is available from the British library

ISBN 1-84236-477-4

Printed in India

Publishing Director: Anne Marshall
Project Manager: Ian Paulyn
Cover Design: Jo Brewer
Production Manager: Estela Boulton
Written and designed by Barnsbury Books

www.mileskelly.net
info@mileskelly.net

4000 MORE FACTS

John Farndon

BARDFIELD
PRESS

Contents

SCIENCE 14-69

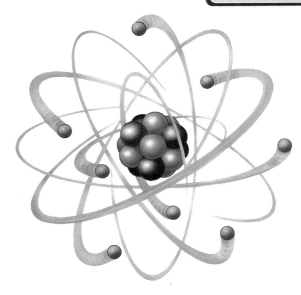

Contents

BUILDINGS & TRANSPORT 70-127

Contents

GEOGRAPHY 128-185

Contents

PLANTS 186-243

INDEX 244-256

INTRODUCTION

This incredible reference resource provides knowledge, fascination and inspiration on every page. Its four hundred subject panels contain facts that will inform, amaze and entertain. You will find out quickly and effortlessly about science, buildings and transport, geography and plants. And you will learn through facts like these:

• *One kilogram of deuterium (a kind of hydrogen) can give the same amount of energy as three million kilograms of coal.*

• *The McLaren F1 can accelerate from 0-160 km/h in less time than it takes to read this sentence.*

• *The Amazon basin is home to 30 million different kinds of insect.*

• *Every year the world uses three billion cubic metres of wood – a pile as big as a football stadium and as high as Mt Everest.*

Now discover the other 3996 brilliant facts in **4000 More Things You Should Know.**

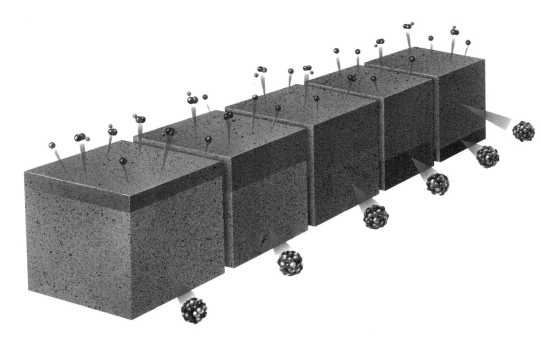

Using this book

The organization of **4000 More Things You Should Know** brings surprises and interest to every page.

The book is divided into four broad areas: Science, Buildings and Transport, Geography and Plants.

On each double-page spread there are three or four subject panels. Each panel contains 10 key facts and is identified with a highlighted subject symbol. You can turn to subjects that interest you by looking for the subject symbols below.

The subjects covered on each spread are organized so that you will find interest and variety throughout the book. Use the subject symbols, the contents page and the index to navigate.

Harvesting grain

- When grain is ripe it is cut from its stalks. This is called reaping.
- After reaping the grain must be separated from the stalks and chaff (waste). This is called threshing.
- After threshing the grain must be cleaned and separated from the husks. This is called winnowing.
- In some places grain is still reaped in the ancient way with a long curved blade called a sickle.
- In most developed countries wheat and other cereals are usually harvested with a combine harvester.
- A combine harvester is a machine that reaps the grain, threshes it, cleans it and pours it into bags or reservoirs.
- The first horse-drawn combine was used in Michigan in 1836, but modern self-propelled harvesters only came into use in the 1940s.
- If the grain is damp it must be dried immediately after harvesting so it does not rot. This is always true of rice.

▲ Combine harvesters driven by a single man have repla... huge teams of people with sickles of ancient times.

- If the grain is too damp to harvest, a machine ca... windrower may cut the stalks and lay them in row... in the wind for later threshing and cleaning.
- A successful harvest is traditionally celebrated w... harvest festival. The cailleac or last sheaf of corn i... be the spirit of the field. It is made into a harvest ... drenched with water and saved for the spring plan...

Desert plants

▲ Surprisingly many plants can survive the dryness of deserts, including cactuses and sagebushes.

- Some plants find water in the dry desert with very long roots. The Mesquite has roots that can go down as much as 50 m deep.
- Most desert plants have tough waxy leaves to cut down on water loss. They also have very few leaves; cactuses have none at all.

- Pebble plants avoid the desert heat by growing p... underground.
- Window plants grow almost entirely undergroun... long cigar shape pokes into the ground, with jus... green window on the surface to catch sunlight.
- Some mosses and lichens get water by soaking ...
- Resurrection trees get their name because their ... look shrivelled brown and dead most of the time – ... suddenly turn green when it rains.
- The rose of Jericho is a resurrection plant that fo... dry ball that lasts for years and opens only when c...
- Daisies are found in most deserts.
- Cactuses and ice plants can store water for m... months in special water storage organs.

★ STAR FACT ★
The quiver tree drops its branches to save... water in times of drought.

Over 300 photographs help illustrate the amazing facts.

SUBJECT SYMBOLS

SCIENCE

 Matter

 Chemicals and materials

Electricity, magnetism and radiation

The frontiers of science

Technology

Energy, force and motion

BUILDINGS AND TRANSPORT

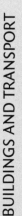 Cars

 Trains

 Planes

 Buildings

 Great monuments

 Boats

GEOGRAPHY

Asia

The Americas

Europe

Africa and Australasia

People

Places

PLANTS

How plants work

Flowers

Biomes

Mosses etc.

Trees

Plants and humans

Subject symbols appear on every panel. Look for the ones that are highlighted.

'Newsflashes' give you up to the minute snippets of information. Star facts are strange-but-true.

There are diagrams throughout the book.

Headings at the top of each double-page spread tell you which of the four areas of the book you are in – Buildings and Transport, Science, Plants or Geography.

Ten key facts are provided in each subject panel. There are 400 panels making 4000 facts in all.

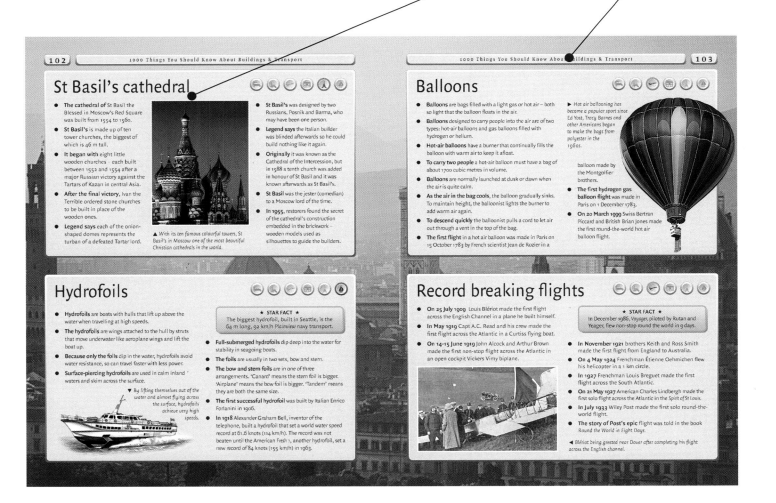

1000 THINGS
YOU SHOULD KNOW ABOUT

SCIENCE

KEY

 Matter

 Chemicals and materials

 Electricity, magnetism and radiation

 The frontiers of science

 Technology

 Energy, force and motion

Computers

- **Part of a computer's** memory is microchips built in at the factory and known as ROM, or read-only memory. ROM carries the basic working instructions.

- **RAM** (random-access memory) consists of microchips that receive new data and instructions when needed.

- **Data can also** be stored as magnetic patterns on a removable disk, or on the laser-guided bumps on a CD (compact disk) or DVD (digital versatile disk).

- **At the heart** of every computer is a powerful

◀ *Computers are developing so rapidly that models from the 1990s already look dated.*

microchip called the central processing unit, or CPU.

- **The CPU** works things out, within the guidelines set by the computer's ROM. It carries out programs by sending data to the right place in the RAM.

- **Computers** store information in bits (binary digits), either as 0 or 1.

- **The bits 0 and 1** are equivalent to the OFF and ON of electric current flow. Eight bits make a byte.

- **A kilobyte** is 1000 bytes; a megabyte (MB) is 1,000,000 bytes; a gigabyte (GB) is 1,000,000,000 bytes; a terabyte (TB) is 1,000,000,000,000 bytes.

- **A CD can hold** about 600 MB of data – about 375,000 pages of ordinary text.

> ★ STAR FACT ★
> The US Library of Congress's 70 million books could be stored in 25 TB of computer capacity.

Turning forces

- **Every force** acts in a straight line. Things move round because of a 'turning effect'.

- **A turning effect** is a force applied to an object that is fixed or pivots in another place, called the fulcrum.

- **In a door** the fulcrum is the hinge.

- **The size of a turning force** is known as the moment.

▲ *Interlocking gear wheels are used in a huge range of machines, from cars to cameras. The wheels transmit movements and control their speed and size.*

- **The farther from the fulcrum** that a force is applied, the bigger the moment is.

- **A lever** makes it much easier to move a load by making use of the moment (size of turning force).

- **A first-class lever,** such as pliers or scissors, has the fulcrum between the effort and the load; a second-class lever, such as a screwdriver or wheelbarrow, has the load between the effort and the fulcrum; a third-class lever, such as your lower arm or tweezers, has the effort between the load and the fulcrum.

- **Gears are sets of wheels** of different sizes that turn together. They make it easier to cycle uphill, or for a car to accelerate from a standstill, by spreading the effort over a greater distance.

- **The gear ratio** is the number of times that the driving wheel turns the driven wheel for one revolution of itself.

- **The larger the gear ratio** the more the turning force is increased, but the slower the driven wheel turns.

Archimedes

- **Archimedes** (c.287–212BC) was one of the first great scientists. He created the sciences of mechanics and hydrostatics.

- **Archimedes** was a Greek who lived in the city of Syracuse, Sicily. His relative, Hieron II, was king of Syracuse.

- **Archimedes' screw** is a pump supposedly invented by Archimedes. It scoops up water with a spiral device that turns inside a tube. It is still used in the Middle East.

- **To help defend** Syracuse against Roman attackers in 215BC, Archimedes invented many war machines. They included an awesome 'claw' – a giant grappling crane that could lift galleys from the water and sink them.

- **Archimedes** was killed by Roman soldiers when collaborators let the Romans into Syracuse in 212BC.

- **Archimedes** analysed levers mathematically. He showed that the load you can move with a particular effort is in exact proportion to its distance from the fulcrum.

- **Archimedes discovered** that things float because they are thrust upwards by the water.

- **Archimedes' principle** shows that the upthrust on a floating object is equal to the weight of the water that the object pushes out of the way.

- **Archimedes** realized he could work out the density, or specific gravity, of an object by comparing its weight to the weight of water it pushes out of a jar when submerged.

- **Archimedes** used specific gravity to prove a goldsmith had not made King Hieron's crown of pure gold.

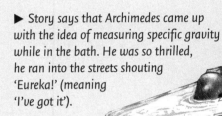

▶ Story says that Archimedes came up with the idea of measuring specific gravity while in the bath. He was so thrilled, he ran into the streets shouting 'Eureka!' (meaning 'I've got it').

Magnetism

- **Magnetism** is the invisible force between materials such as iron and nickel. Magnetism attracts or repels.

- **A magnetic field** is the area around a magnet inside which its magnetic force can be detected.

- **An electric current** creates its own magnetic field.

- **A magnet** has two poles: a north pole and a south pole.

- **Like (similar) poles** (e.g. two north poles) repel each other; unlike poles attract each other.

★ STAR FACT ★
One of the world's strongest magnets is at the Lawrence Berkeley National Laboratory, California, USA. Its field is 250,000 times stronger than the Earth's.

- **The Earth** has a magnetic field that is created by electric currents inside its iron core. The magnetic north pole is close to the geographic North Pole.

- **If left to swivel freely**, a magnet will turn so that its north pole points to the Earth's magnetic north pole.

- **The strength of a magnet** is measured in teslas. The Earth's magnetic field is 0.00005 teslas.

- **All magnetic materials** are made up of tiny groups of atoms called domains. Each one is like a mini-magnet with north and south poles.

◀ The glowing skies above the poles are called auroras. They are created by the way the Earth's magnetic field channels electrically charged particles from the Sun down into the atmosphere.

New materials

- **Synthetic materials** are man-made, such as plastics.

- **Many synthetic materials** are polymers. These are substances with long chains of organic molecules made up from lots of identical smaller molecules, monomers.

- **Some polymers** are natural, such as the plant fibre cellulose.

▲ Snowboards are made from composites such as Kevlar, which combine lightness with strength.

- **The first synthetic polymer** was Parkesine, invented by Alexander Parkes in 1862. The first successful synthetic polymer was celluloid, invented by John Hyatt in 1869 and soon used for photographic film.

- **Nylon** (a polymer) was the first fully synthetic fibre, created by Wallace Carothers of Du Pont in the 1930s.

- **PVC** is polyvinyl chloride, a synthetic polymer developed in the 1920s.

- **Composites** are new, strong, light materials created by combining a polymer with another material.

- **Carbon-reinforced plastic** consists of tough carbon fibres set within a polymer.

- **Kevlar** is a composite made by Du Pont in 1971. It is made of aramid (nylon-like) fibres set within a polymer.

Soaps

- **Some soaps** are natural; all detergents are synthetic.

- **All soaps and detergents** clean with a 'surfactant'.

- **Surfactants** are molecules that attach themselves to particles of dirt on dirty surfaces and lift them away.

- **Surfactants** work because one part of them is hydrophilic (attracted to water) and the other is hydrophobic (repelled by water).

- **The hydrophobic tail** of a surfactant digs its way into the dirt; the other tail is drawn into the water.

- **Soaps** increase water's ability to make things wet by reducing the surface tension of the water.

- **Soap** is made from animal fats or vegetable oil combined with chemicals called alkalis, such as sodium or potassium hydroxide.

- **Most soaps** include perfumes, colours and germicides (germ-killers) as well as a surfactant.

- **The Romans used** soap over 2000 years ago.

- **Detergents** were invented in 1916 by a German chemist called Fritz Gunther.

▼ Surfactant molecules in soap lift dirt off dirty surfaces.

The hydrophobic tail dips into the dirt

The hydrophilic tail is pulled by the water

The surfactant molecules in soap lift particles of dirt away

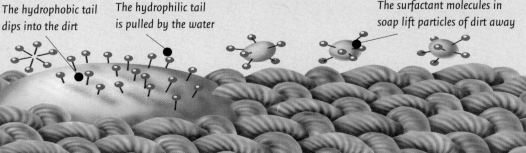

Musical sound

◄ In stringed instruments different notes – that is, different frequencies of vibrations – are achieved by varying the length of the strings.

- **Like all sounds,** musical sounds are made by something vibrating. However, the vibrations of music occur at very regular intervals.

- **The pitch** of a musical note depends on the frequency of the vibrations.

- **Sound frequency** is measured in hertz (Hz) – that is, cycles or waves per second.

> ★ STAR FACT ★
> A song can shatter glass if the pitch of a loud note coincides with the natural frequency of vibration of the glass.

- **Human ears** can hear sounds as low as about 20 Hz and up to around 20,000 Hz.

- **Middle C** on a piano measures 262 Hz. A piano has a frequency range from 27.5 to 4186 Hz.

- **The highest singing voice** can reach the E above a piano top note (4350 Hz); the lowest note is 20.6 Hz.

- **A soprano's voice** has a range from 262 to 1046 Hz; an alto from 196 to 698 Hz; a tenor from 147 to 466 Hz; a baritone from 110 to 392 Hz; a bass from 82.4 to 294 Hz.

- **Very few sounds** have only one pitch. Most have a fundamental (low) pitch and higher overtones.

- **The science of vibrating strings** was first worked out by Pythagoras 2500 years ago.

▶ In most brass and woodwind instruments, such as a tuba, different frequencies are achieved by varying the length of the air column inside.

Lasers

- **Laser light** is a bright artificial light. It creates an intense beam that can punch a hole in steel. A laser beam is so straight and narrow that it can hit a mirror on the Moon.

- **The name** 'laser' stands for **l**ight **a**mplification by **s**timulated **e**mission of **r**adiation.

- **Laser light** is even brighter for its size than the Sun.

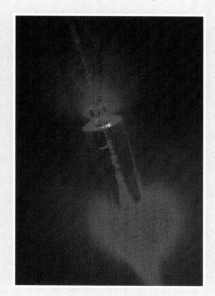

▲ The amazingly tight intense beam of a laser is used in a huge number of devices, from CD players to satellite guidance systems.

- **Laser light** is the only known 'coherent' source of light. This means the light waves are not only all the same wavelength (colour), but they are also perfectly in step.

- **Inside a laser** is a tube filled with gases, such as helium and neon, or a liquid or solid crystal such as ruby.

- **Lasers work** by bouncing photons (bursts of light) up and down the tube until they are all travelling together.

- **Lasing begins** when a spark excites atoms in the lasing material. The excited atoms emit photons. When the photons hit other atoms, they fire off photons too. Identical photons bounce backwards and forwards between mirrors at either end of the laser.

- **Gas lasers** such as argon lasers give a lower-powered beam. It is suitable for delicate work such as eye surgery.

- **Chemical lasers** use liquid hydrogen fluoride to make intense beams for weapons.

- **Some lasers** send out a continuous beam. Pulsed lasers send out a high-powered beam at regular intervals.

Aluminium

- **Aluminium** is by far the most common metal on the Earth's surface. It makes up 8% of the Earth's crust.

- **Aluminium** never occurs naturally in its pure form; in the ground it combines with other chemicals as minerals in ore rocks.

- **The major source** of aluminium is layers of soft ore called bauxite, which is mostly aluminium hydroxide.

- **Alum powders** made from aluminium compounds were used 5000 years ago for dyeing. Pure aluminium was first made in 1825 by Danish scientist Hans Oersted.

- **Aluminium** production was the first industrial process to use hydroelectricity when a plant was set up on the river Rhine in 1887.

- **Aluminium is silver** in colour when freshly made, but it quickly

▲ *Although aluminium is common in the ground, it is worth recycling because extracting it from bauxite uses a lot of energy.*

tarnishes to white in the air. It is very slow to corrode.

- **Aluminium** is one of the lightest of all metals.

- **Aluminium oxide** can crystallize into one of the hardest minerals, corundum, which is used to sharpen knives.

- **Aluminium** melts at 650°C and boils at 2450°C.

- **Each year 21 million tonnes** of aluminium are made, mostly from bauxite dug up in Brazil and New Guinea.

◄ *Half of the soft drinks cans in the USA are made from recycled aluminium.*

Stretching and pulling

▲ *The leverage of the bow string helps an archer to bend the elastic material of the bow so far that it has tremendous power as it snaps back into shape.*

- **Elasticity** is the ability of a solid material to return to its original shape after it has been misshapen.

- **A force** that misshapes material is called a stress.

- **All solids** are slightly elastic but some are very elastic, for example rubber, thin steel and young skin.

- **Solids** return to their original shape after the stress stops, as long as the stress is less than their 'elastic limit'.

- **Strain** is how much a solid is stretched or squeezed when under stress, namely how much longer it grows.

- **The amount** a solid stretches under a particular force – the ratio of stress to strain – is called the elastic modulus, or Young's modulus.

- **The greater the stress**, the greater the strain. This is called Hooke's law, after Robert Hooke (1635–1703).

- **Solids** with a low elastic modulus, like rubber, are stretchier than ones with a high modulus, such as steel.

- **Steel can be only stretched** by 1% before it reaches its elastic limit. If the steel is coiled into a spring, this 1% can allow a huge amount of stretching and squeezing.

> ★ **STAR FACT** ★
> Some types of rubber can be stretched 1000 times beyond its original length before it reaches its elastic limit.

Electrons

- **Electrons** are by far the smallest of the three main, stable parts of every atom; the other two parts are protons and neutrons (see atoms). In a normal atom there are the same number of electrons as protons.

- **Electrons** are 1836 times as small as protons and have a mass of just 9.109×10^{-31} kg. 10^{-31} means there are 30 zeros after the decimal point. So they weigh almost nothing.

- **Electrons were discovered** by English physicist Joseph John Thomson in 1897 as he studied the glow in a cathode-ray tube (see television). This was the first a anyone realized that the atom is not just one solid ball.

- **Electrons are** packets of energy. They can be thought of either as a tiny vibration or wave, or as a ball-like particle. They travel as waves and arrive as particles.

- **You can never be sure** just where an electron is. It is better to think of an electron circling the nucleus not as a planet circling the Sun but as a cloud wrapped around it. Electron clouds near the nucleus are round, but those farther out are other shapes, such as dumb-bells.

- **Electrons** have a negative electrical charge. This means they are attracted to positive electrical charges and pushed away by negative charges. 'Electron' comes from the Greek word for amber. Amber tingles electrically when rubbed.

- **Electrons cling** to the nucleus because protons have a positive charge equal to the electron's negative charge.

▶ *Each atom has a different number of electrons. Its chemical character depends on the number of electrons in its outer shell. Atoms with only one electron in their outer shell, such as lithium, sodium and potassium, have many properties in common. The electron shell structures for five common atoms are shown here.*

★ **STAR FACT** ★
Electrons whizz round an atomic nucleus, zoom through an electric wire or spin on their own axis, either clockwise or anti-clockwise.

- **Electrons have so much energy** that they whizz round too fast to fall into the nucleus. Instead they circle the nucleus in shells (layers) at different distances, or energy levels, depending on how much energy they have. The more energetic an electron, the farther from the nucleus it is. There is room for only one other electron at each energy level, and it must be spinning in the opposite way. This is called Pauli's exclusion principle.

- **Electrons are** stacked around the nucleus in shells, like the layers of an onion. Each shell is labelled with a letter and can hold a particular number of electrons. Shell K can hold up to 2, L 8, M 18, N 32, O about 50, and P about 72.

Hydrogen atom

Single electron

Nucleus with single proton

Carbon atom

Nucleus with 6 protons

Maximum 2 electrons in shell K

Shell L holds 4 electrons out of a possible 8. So carbon has four vacancies to form complex compounds with other elements.

Oxygen atom

Nucleus with 8 protons

Shell K holds a maximum of 2 electrons.

Shell L holds 6 electrons out of a possible 8. So oxygen has 2 'missing' electrons and is very reactive.

Shell L holds a maximum of 8 electrons, so the next electron goes in shell M.

Single electron in shell M is easily drawn to other atoms.

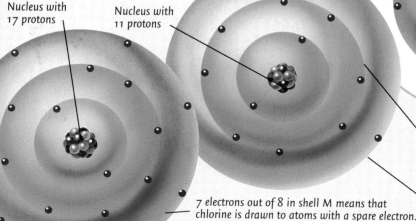

Chlorine atom

Nucleus with 17 protons

Sodium atom

Nucleus with 11 protons

7 electrons out of 8 in shell M means that chlorine is drawn to atoms with a spare electron.

Oxygen

▲ *No animal can live for more than a minute or so without breathing in oxygen from the air to keep the body processes going.*

- **Oxygen** is the second most plentiful element on Earth. Air is 20.94% oxygen.

- **Oxygen** is one of the most reactive elements. This is why oxygen in the Earth's crust is usually found joined with other chemicals in compounds.

- **Oxygen has an atomic number** of 8 and an atomic weight of 15.9994.

- **Oxygen molecules** in the air are made from two oxygen atoms; three oxygen atoms make the gas ozone.

- **Oxygen turns to a pale blue liquid** at −182.962°C. It freezes at −218.4°C.

- **Most life depends on oxygen** because it joins with other chemicals in living cells to give the energy needed for life processes. The process of using oxygen in living cells is called cellular respiration.

- **Liquid oxygen,** or LOX, is combined with fuels such as kerosene to provide rocket fuel.

- **Oxygen** was discovered independently by Carl Scheele and Joseph Priestley during the 1770s.

- **The name** 'oxygen' means acid-forming. It was given to the gas in 1779 by Antoine Lavoisier (see Lavoisier).

> ★ STAR FACT ★
> The oxygen in the air on which your life depends was produced mainly by algae.

Einstein

▶ *Einstein's equation E=mc² revealed the energy in atoms that led to nuclear bombs and nuclear power.*

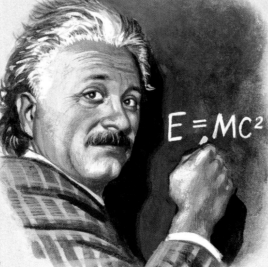

- **Albert Einstein** (1879–1955) was the most famous scientist of the 20th century.

- **Einstein was half German** and half Swiss, but when Hitler came to power in 1933, Einstein made his home in the USA.

- **Einstein's fame** rests on his two theories of Relativity (see relativity).

- **His theory of Special Relativity** was published in 1905 while he worked in the Patent Office in Bern, Switzerland.

- **In 1905** Einstein also explained the photoelectric effect. From these ideas, photo cells were developed. These are the basis of TV cameras and many other devices.

- **Einstein completed his theory** of General Relativity in 1915.

- **Einstein** was not satisfied with his theory of General Relativity as it didn't include electromagnetism. He spent the last 25 years of his life trying to develop a 'Unified Field Theory' to include it.

- **Einstein** was once reported to have said that only 12 people in the world could understand his theory. He denied saying it.

- **On August 2, 1939** Einstein wrote a letter to US President Franklin Roosevelt in which he persuaded the president to launch the Manhattan Project to develop the nuclear bomb.

- **Einstein was married twice.** His first wife was Mileva Maric. His second wife Elsa was also his first cousin.

Atoms

- **Atoms are** the tiny bits, or particles, which build together to make every substance. An atom is the tiniest bit of any pure substance or chemical element.

- **You could fit** two billion atoms on the full stop after this sentence.

- **The number of atoms** in the Universe is about 10 followed by 80 zeros.

- **Atoms are mostly** empty space dotted with a few even tinier particles called subatomic particles.

- **In the centre** of each atom is a dense core, or nucleus, made from two kinds of particle: protons and neutrons. Protons have a positive electrical charge, and neutrons none. Both protons and neutrons are made from different combinations of quarks (see quarks).

- **If an atom** were the size of a sports arena, its nucleus

◀ Inside every atom tiny electrons (blue) whizz around a dense nucleus built up from protons (red) and neutrons (green). The numbers of each particle vary from element to element.

would be the size of a pea.

- **Around the nucleus** whizz even tinier, negatively-charged particles called electrons (see electrons).

- **Atoms can be split** but they are usually held together by three forces: the electrical attraction between positive protons and negative electrons, and the strong and weak 'nuclear' forces that hold the nucleus together.

- **Every element** is made from atoms with a certain number of protons in the nucleus. An iron atom has 26 protons, gold has 79. The number of protons is the atomic number.

- **Atoms with the same number** of protons but a different number of neutrons are called isotopes.

Microscopes

- **Microscopes** are devices for looking at things that are normally too small for the human eye to see.

- **Optical microscopes** use lenses to magnify the image by up to 2000 times.

- **In an optical microscope** an objective lens bends light rays apart to enlarge what you see; an eyepiece lens makes the big image visible.

- **Electron microscopes** magnify by firing electrons at an object. The electrons bounce off the object onto a screen, making them visible.

- **An electron microscope** can focus on something as small as 1 nanometre (one-billionth of a millimetre) and magnify it five million times.

- **Scanning Electron Microscopes** (SEMs) scan the surface of an object to magnify it by up to 100,000 times.

- **Transmission Electron Microscopes** shine electrons through thin slices of an object to magnify it millions of times.

- **Scanning Tunnelling Microscopes** are so powerful that they can reveal individual atoms.

- **The idea of electron microscopes** came from French physicist Louis de Broglie in 1924.

- **Scanning Acoustic Microscopes** use sound waves to see inside tiny opaque objects.

◀▼ A Scanning Electron Microscope clearly reveals the tiny nerve fibres inside the human brain.

Electromagnetic spectrum

- **The electromagnetic spectrum** is the complete range of radiation sent out by electrons (see light and atoms). It is given off in tiny packages of energy called photons, which can be either particles or waves (see moving light).

- **Electromagnetic waves** vary in length and frequency. The shorter the wave, the higher its frequency (and also its energy).

- **The longest waves** are over 100 kilometres long; the shortest are less than a billionth of a millimetre long.

- **All electromagnetic waves** travel at 300,000 kilometres per second, which is the speed of light.

- **Visible light** is just a small part of the spectrum.

- **Radio waves,** including microwaves and television waves, and infrared light, are made from waves that are too long for human eyes to see.

- **Long waves** are lower in energy than short waves. Long waves from space penetrate the Earth's atmosphere easily (but not solids, like short waves).

- **Ultraviolet light,** X-rays and gamma rays are made from waves that are too short for human eyes to see.

- **Short waves are very energetic.** But short waves from space are blocked out by Earth's atmosphere – which is fortunate because they are dangerous. X-rays and gamma rays penetrate some solids, and UV rays can damage living tissues, causing skin cancers.

> ★ STAR FACT ★
> Cosmic rays are not rays but streams of high-energy particles from space.

Gamma rays are dangerous high-energy rays with such short waves that they can penetrate solids. They are created in space and by nuclear bombs.

X-rays are longer waves than gamma rays but short enough to pass through most body tissues except bones, which show up white on medical X-ray photos.

The shortest ultraviolet rays in sunshine are dangerous, but longer ones give you a suntan in small doses. In large doses even long UV rays cause cancer.

Visible light varies in wavelength from violet (shortest) through all the colours of the rainbow to red (longest).

◀ *This illustration shows the range of radiation in the electromagnetic spectrum. The waves are shown emerging from the Sun because the Sun actually emits almost the full range of radiation. Fortunately, the atmosphere protects us from the dangerous ones.*

Infrared light is the radiation given out by hot objects. This is why infrared-sensitive 'thermal imaging' cameras can see hot objects such as people in pitch darkness.

Microwaves are used to beam telephone signals to satellites – and to cook food. Radars send out fairly short microwaves (about 1 cm long).

Television broadcasts use radio waves with waves about 0.5 m long.

Radio broadcasts use radio waves with waves from 300 m to 1500 m long.

Typical wavelength in metres or millimetres. Long waves are low frequency and low energy. Short waves are high frequency and high energy.

1 billionth mm

10 millionth mm

0.00001 mm

0.0005 mm

0.2 mm

0.3 m to 1 mm

0.5 m

300–1500 m

Chemical bonds

- **Chemical bonds** link together atoms to make molecules (see molecules).

- **Atoms can bond** in three main ways: ionic bonds, covalent bonds and metallic bonds.

- **In ionic bonds** electrons are transferred between atoms.

- **Ionic bonds** occur when atoms with just a few electrons in their outer shell give the electrons to atoms with just a few missing from their outer shell.

- **An atom** that loses an electron becomes positively charged; an atom that gains an electron becomes negatively charged so the two atoms are drawn together by the electrical attraction of opposites.

- **Sodium** loses an electron and chlorine gains one to form the ionic bond

◄ *Each of the four hydrogen atoms in methane (CH$_4$) shares its electron with the central carbon atom to create strong covalent bonds.*

▶ *In this carbon dioxide molecule the carbon is held to two oxygen atoms by covalent bonds.*

of sodium chloride (table salt) molecules.

- **In covalent bonding,** the atoms in a molecule share electrons.

- **Because they are negatively** charged, the shared electrons are drawn equally to the positive nucleus of both atoms involved. The atoms are held together by the attraction between each nucleus and the shared electrons.

- **In metallic bonds** huge numbers of atoms lose their electrons. They are held together in a lattice by the attraction between 'free' electrons and positive nuclei.

> **★ STAR FACT ★**
> Seven elements, including hydrogen, are found in nature only as two atoms covalently bonded.

Pressure

▶ *The worst storms, such as this hurricane seen from space, are caused when air from high-pressure areas rushes into low-pressure areas*

- **Pressure** is the force created by the assault of fast-moving molecules.

- **The pressure that keeps** a bicycle tyre inflated is the constant assault of huge numbers of air molecules on the inside of the tyre.

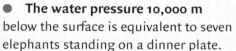

- **The water pressure that** crushes a submarine when it dives too deep is the assault of huge numbers of water molecules.

- **Pressure rises** as you go deeper in the ocean. This is because of an increasing weight of water – called hydrostatic pressure – pressing down from above.

- **The water pressure 10,000 m** below the surface is equivalent to seven elephants standing on a dinner plate.

- **The pressure of the air** on the outside of your body is balanced by the pressure of fluids inside. Without this internal pressure, air pressure would crush your body instantly.

- **Pressure** is measured as the force on a certain area.

- **The standard unit** of pressure is a pascal (Pa) or 1 newton per sq m (N/m^2).

- **High pressures:** the centre of the Earth may be 400 billion Pa; steel can withstand 40 million Pa; a shark bite can be 30 million Pa.

- **Low pressures:** the best laboratory vacuum is 1 trillionth Pa; the quietest sound is 200 millionths Pa. The pressure of sunlight may be 3 millionths Pa.

Machines

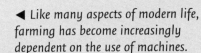

- **A machine** is a device that makes doing work easier by reducing the Effort needed to move something.

- **There are two forces** involved in every machine: the Load that the machine has to overcome, and the Effort used to move the Load.

- **The amount that a machine** reduces the Effort needed to move a Load is called the Mechanical Advantage. This tells you how effective a machine is.

- **Basic machines include** levers, gears, pulleys, screws, wedges and wheels. More elaborate machines, such as cranes, are built up from combinations of these basic machines.

- **Machines cut** the Effort needed to move a Load by spreading the Effort over a greater distance or time.

- **The distance** moved by the Effort you apply, divided

◀ *Like many aspects of modern life, farming has become increasingly dependent on the use of machines.*

by the distance moved by the Load, is called the Velocity Ratio (VR).

- **If the VR is greater** than 1, then the Effort moves farther than the Load. You need less Effort to move the Load, but you have to apply the Effort for longer.

- **The total amount** of Effort you use to move something is called Work. Work is the force you apply multiplied by the distance that the Load moves.

- **One of the earliest machines still used today** is a screwlike water-lifting device called a dalu, first used in Sumeria 5500 years ago.

- **One of the world's biggest machines** is the SMEC earthmover used in opencast mines in Australia. It weighs 180 tonnes and has wheels 3.5 m high.

Faraday

▲ *Faraday drew huge crowds to his brilliant and entertaining Christmas lectures on science at the Royal Institution in London. These Christmas lectures at the Royal Institution are still a popular tradition today.*

- **Michael Faraday** (1791–1867) was one of the greatest scientists of the 19th century.

- **Faraday was the son** of a poor blacksmith, born in the village of Newington in Surrey, England.

- **He started work as** an apprentice bookbinder but became assistant to the great scientist Humphry Davy after taking brilliant notes at one of Davy's lectures.

- **Faraday was said** to be Davy's greatest discovery.

- **Until 1830 Faraday** was mainly a chemist. In 1825 he discovered the important chemical benzene.

- **In 1821** Faraday showed that the magnetism created by an electric current would make a magnet move and so made a very simple version of an electric motor.

- **In 1831** Faraday showed that when a magnet moves close to an electric wire, it creates, or induces, an electric current in the wire. This was discovered at the same time by Joseph Henry in the USA.

- **Using** his discovery of electric induction, Faraday made the first dynamo to generate electricity and so opened the way to the modern age of electricity.

- **In the 1840s** Faraday suggested the idea of lines of magnetic force and electromagnetic fields. These ideas, which were later developed by James Clerk Maxwell, underpin much of modern science.

- **Faraday** was probably the greatest scientific experimenter of all time.

Electric currents

- **An electric charge** that does not move is called static electricity (see electricity). A charge may flow in a current providing there is an unbroken loop, or circuit.

- **A current only flows** through a good conductor such as copper, namely a material that transmits charge well.

- **A current only flows** if there is a driving force to push the charge. This force is called an electromotive force (emf).

- **The emf** is created by a battery or a generator.

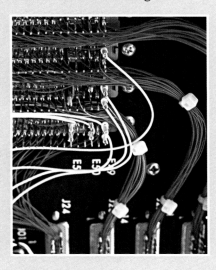

▶ For a current to flow, each of these wires must form a circuit.

★ STAR FACT ★
The electrical resistance of dry skin is 500,000 ohms; wet skin's is just 1000 ohms.

- **Currents were once thought to** flow like water. In fact they move like a row of marbles knocking into each other.

- **In a good conductor** there are lots of free electrons that are unattached to atoms. These are the 'marbles'.

- **A current only flows** if there are more electrons at one point in the circuit. This difference, called the potential difference, is measured in volts.

- **The rate at which current** flows is measured in amps. It depends on the voltage and the resistance (how much the circuit obstructs the flow of current). Resistance is measured in ohms.

- **Batteries** give out Direct Current, a current that flows in one direction. Power stations send out Alternating Current, which swaps direction 50–60 times per second.

Plastics

- **Plastics are synthetic** (man-made) materials that can be easily shaped and moulded.

- **Most plastics are polymers** (see new materials). The structure of polymer molecules gives different plastics different properties.

- **Long chains of molecules** that slide over each other easily make flexible plastics such as polythene. Tangled chains make rigid plastics such as melamine.

- **Typically** plastics are made by joining carbon and hydrogen atoms. These form ethene molecules, which can be joined to make a plastic called polythene.

- **Many plastics** are made from liquids and gases that are extracted from crude oil.

- **Thermoplastics** are soft and easily moulded when warm but set solid when cool. They are used to make bottles and drainpipes and can be melted again.

- **Thermoset plastics,** which cannot be remelted once set, are used to make telephones and pan handles.

- **Blow moulding** involves using compressed air to push a tube of plastic into a mould.

- **Vacuum moulding** involves using a vacuum to suck a sheet of plastic into a mould.

- **Extrusion moulding** involves heating plastic pellets and forcing them out through a nozzle to give the right shape.

▼ Some plastics are light and soft and can be filled with air bubbles to make an ideal packing material.

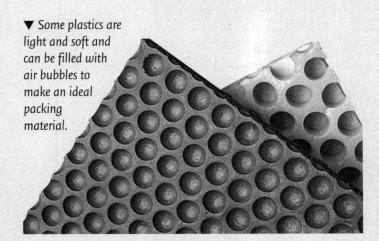

Chemical compounds

- **Compounds** are substances that are made when the atoms of two or more different elements join together.

- **The properties of a compound** are usually very different from those of the elements which it is made of.

- **Compounds** are different from mixtures because the elements are joined together chemically. They can only be separated by a chemical reaction.

- **Every molecule** of a compound is exactly the same combination of atoms.

- **The scientific name** of a compound is usually a combination of the elements involved, although it might have a different common name.

- **Table salt** is the chemical compound sodium chloride. Each molecule has one sodium and one chlorine atom.

- **The chemical formula** of a compound summarizes which atoms a molecule is made of. The chemical formula for water is H_2O because each water molecule has two hydrogen (H) atoms and one oxygen (O) atom.

- **There only 100 or so elements** but they can combine in different ways to form many millions of compounds.

- **The same combination of elements,** such as carbon and hydrogen, can form many different compounds.

- **Compounds** are either organic (see organic chemistry), which means they contain carbon atoms, or inorganic.

▼ *The molecules of a compound are identical combinations of atoms.*

Halogens

▲ *Chlorine salts are often added to the water in swimming pools to kill bacteria, giving the water a greenish-blue tinge.*

- **Halogens** are the chemical elements fluorine, chlorine, bromine, iodine and astatine.

- **The word 'halogen'** means salt-forming. All halogens easily form salt compounds.

- **Many of the salts in the sea** are compounds of a halogen and a metal, such as sodium chloride and magnesium chloride.

- **The halogens** all have a strong, often nasty, smell.

- **Fluorine** is a pale yellow gas, chlorine a greenish gas, bromine a red liquid, and iodine a black solid.

- **Astatine** is an unstable element that survives by itself only briefly. It is usually made artificially.

- **The halogens** together form Group 17 of the Periodic Table, elements with 7 electrons in the outer shells.

- **Because halogens have** one electron missing, they form negative ions and are highly reactive.

- **The iodine and bromine** in a halogen lightbulb make it burn brighter and longer.

★ **STAR FACT** ★
Fluorides (fluorine compounds) are often added to drinking water to prevent tooth decay.

Particle physics

- **Apart from the three** basic, stable particles of atoms – electrons, protons and neutrons – scientists have found over 200 rare or short-lived particles. Some were found in cosmic rays from space; some appear when atoms are smashed to bits in devices called particle accelerators.

- **Every particle** also has a mirror-image anti-particle. Although Antimatter may be rarer, it is every bit as real.

- **Cosmic rays** contain not only electrons, protons and neutrons, but short-lived particles such as muons and strange quarks. Muons flash into existence for 2.2 micro-seconds just before the cosmic rays reach the ground.

- **Smashing atoms** in particle accelerators creates short-lived high-energy particles such as taus and pions and three kinds of quark called charm, bottom and top.

- **Particles** are now grouped into a simple framework called the Standard Model. It divides them into elementary particles and composite particles to stop radiation leaks.

- **Elementary particles** are the basic particles which cannot be broken down in anything smaller. There are three groups: quarks, leptons and bosons. Leptons include electrons, muons, taus and neutrinos. Bosons are 'messenger' particles that link the others. They include photons and gluons which 'glue' quarks together.

- **Composite particles** are hadrons made of quarks glued together by gluons. They include protons, neutrons and 'hyperons' and 'resonances'.

> ★ **STAR FACT** ★
> When the Fermi particle accelerator is running 250,000 particle collisions occur every second.

- **To smash atoms** scientists use particle accelerators, which are giant machines set in tunnels. The accelerators use powerful magnets to accelerate particles through a tube at huge speeds, and then smash them together.

- **Huge detectors** pick up collisions between particles.

▼ A view of the particle accelerator tunnel at CERN in Switzerland.

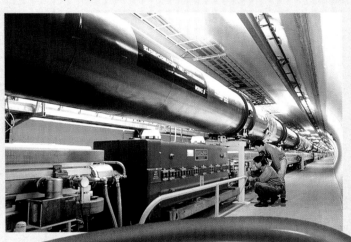

The particles are split up and fed towards the detector from opposite directions so they collide head-on.

Incredibly powerful electromagnets accelerate the particles.

Extra electromagnets keep the particles on track through the pipe.

Some accelerators are ring-shaped so that the particles can whizz round again and again to build up speed.

New particles are fed in from a hot filament like a giant lightbulb filament.

The detectors that record the collisions are like giant electronic cameras. They can be three storeys high and weigh over 5000 tonnes.

The pipes are heavily insulated to stop particles escaping.

◄ The accelerators at Fermilab near Chicago, USA and CERN in Switzerland are underground tubes many kilometres long through which particles are accelerated to near the speed of light.

Forces

- **A force** is a push or a pull. It can make something start to move, slow down or speed up, change direction or change shape or size. The greater a force, the more effect it has.

- **The wind is a force.** Biting, twisting, stretching, lifting and many other actions are also forces. Every time something happens, a force is involved.

- **Force is measured** in newtons (N). One newton is the force needed to speed up a mass of one kilogram by one metre per second every second.

- **When something moves** there are usually several forces involved. When you throw a ball, the force of your throw hurls it forwards, the force of gravity pulls it down and the force of air resistance slows it down.

- **The direction and speed** of movement depend on the combined effect of all the forces involved – this is called the resultant.

- **A force** has magnitude (size) and works in a particular direction.

- **A force can** be drawn on a diagram as an arrow called a vector (see vectors). The arrow's direction shows the force's direction. Its length shows the force's strength.

- **Four fundamental forces** operate throughout the Universe: gravity, electric and magnetic forces (together called electromagnetic force), and strong and weak nuclear forces (see nuclear energy).

- **A force field** is the area over which a force has an effect. The field is strongest closest to the source and gets weaker farther away.

> ★ STAR FACT ★
> The thrust of *Saturn V*'s rocket engines was 33 million newtons.

◄ *When a spacecraft lifts off, the force of the rocket has to overcome the forces of gravity and air resistance to power it upwards.*

Electricity

- **Electricity** is the energy that makes everything from toasters to televisions work. It is also linked to magnetism. Together, as electromagnetism, they are one of the four fundamental forces holding the Universe together.

- **Electricity** is made by tiny bits of atoms called electrons. Electrons have an electrical charge which is a force that either pulls bits of atoms together or pushes them apart.

- **Some particles** (bit of atoms) have a negative electrical charge; others have a positive charge.

- **Particles** with the same charge push each other away. Particles with the opposite charge pull together.

- **Electrons** have a negative electrical charge.

- **There are the same number** of positive and negative particles in most atoms so the charges usually balance.

- **Electricity** is created when electrons move, building up negative charge in one place or carrying it along.

- **Static electricity** is when the negative charge stays in one place. Current electricity is when the charge moves.

- **Electric charge** is measured with an electroscope.

- **Materials** that let electrons (and electrical charge) move through them easily, such as copper, are called conductors. Materials that stop electrons passing through, such as rubber, are called insulators.

▼ *Lightning is a dramatic display of natural electricity.*

Time

▲ As the Sun moves through the sky we can clearly see the passing of time. Until 1967 the movement of the Sun was our most accurate way of measuring time.

- **No clock** keeps perfect time. For most of history clocks were set by the Sun and stars.
- **Since 1967** the world's time has been set by atomic clocks.
- **Atomic clocks** are accurate to 0.001 sec in 1000 years.
- **If a caesium atomic** clock ran for six million years it would not gain or lose a second.
- **The most** accurate clock is the American NIST-7 atomic clock.

- **The atomic clock** on the International Space Station is hundreds of times more accurate than clocks on Earth, because it is not affected by gravity.
- **Atomic clocks** work because caesium atoms vibrate exactly 9,192,631,770 times a second.
- **Some scientists** say that time is the fourth dimension – the other three are length, breadth and height. So time could theoretically run in any direction. Others say time only moves in one direction. Just as we cannot unburn a candle, so we cannot turn back time (see time travel).
- **Light takes** millions of years to reach us from distant galaxies, so we see them not as they are but as they were millions of years ago. Light takes a little while to reach us even from nearby things.
- **Einstein's theory of General Relativity** shows that time actually runs slower nearer strong gravitational fields such as stars. This does not mean that the clock is running slower but that time itself is running slower.

Chemical reactions

- **A candle burning,** a nail rusting, a cake cooking – all involve chemical reactions.
- **A chemical reaction** is when two or more elements or compounds meet and interact to form new compounds or separate out some of the elements.
- **The chemicals** in a chemical reaction are called the reactants. The results are called the products.
- **The products** contain the same atoms as the reactants but in different combinations.
- **The products** have the same total mass as the reactants.
- **Some reactions** are reversible – the products can be changed back to the original reactants. Others, such as making toast, are irreversible.
- **Effervescence** is a reaction in which gas bubbles form in a liquid, turning it fizzy.
- **A catalyst** is a substance that speeds up or enables a chemical reaction to happen, but slows down or remains unchanged at the end.

- **Nearly all reactions** involve energy. Some involve light or electricity. Most involve heat. Reactions that give out heat are called exothermic. Those that draw in heat are called endothermic.
- **Oxidation** is a reaction in which oxygen combines with a substance. Burning is oxidation; as the fuel burns it combines with oxygen in the air. Reduction is a reaction in which a substance loses oxygen.

▶ Burning is an oxidation reaction. Carbon in the trees combines with oxygen in the air to form carbon dioxide.

The Periodic Table

- **The Periodic Table** is a chart of all the 100-plus different chemical elements (see elements).

- **The Periodic Table** was devised by Russian chemist Dmitri Mendeleyev (1834–1907) in 1869. Mendeleyev realized that each element is part of a complete set, and so he predicted the existence of three then unknown elements – gallium, scandium and germanium.

- **The Periodic Table** arranges all the elements according to their Atomic Number, which is the number of protons in their atoms (see atoms). The table lists the elements in order of Atomic Number, starting with hydrogen at 1.

- **Atoms** usually have the same number of electrons (see electrons) as protons. So the Atomic Number also indicates the normal number of electrons an atom has.

- **Atomic mass** is the average weight of an atom of an element and corresponds to the average number of protons and neutrons in the nucleus. The number of neutrons varies in a small number of atoms, called isotopes, so the atomic mass is never a round number.

- **The columns** in the Periodic Table are called Groups. The rows are called Periods.

- **The number** of layers, or shells, of electrons in the atoms of an element increases one by one down each Group. So all the elements in each Period have the same number of electron shells.

- **The number of electrons** in the atom's outer shell goes up one by one across each Period.

- **Each Group** is made up of elements with a certain number of electrons in their outer shell. This is what largely determines the element's character. All the elements in each Group have similar properties. Many of the Groups have a name as well as a number, as below.

- **Each Period** starts on the left with a highly reactive alkali metal of Group 1, such as sodium. Each atom of elements in Group 1 has a single electron in its outer shell. Each Period ends on the right with a stable 'noble' gas of Group 18, such as argon. These elements have the full number of electrons in their outer shell and so they do not react.

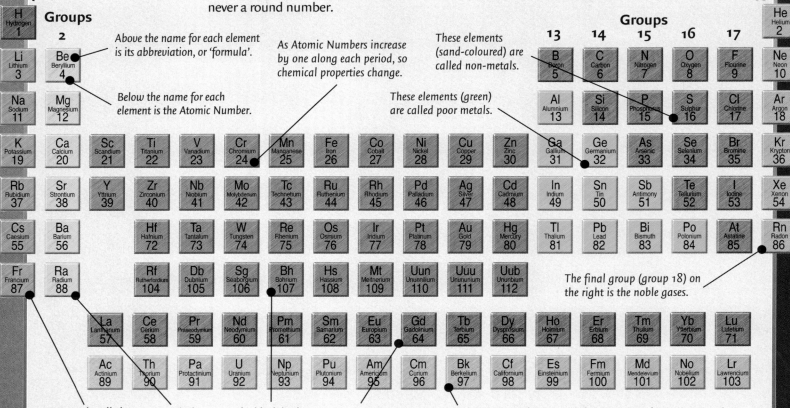

Above the name for each element is its abbreviation, or 'formula'.

Below the name for each element is the Atomic Number.

As Atomic Numbers increase by one along each period, so chemical properties change.

These elements (sand-coloured) are called non-metals.

These elements (green) are called poor metals.

The final group (group 18) on the right is the noble gases.

Group 1 is called alkali metals. They are all soft, very reactive metals.

Group 2 is the alkaline–earth group. They occur naturally only in compounds.

The block in the middle (purple) is transition metals such as gold and copper.

This row, called the lanthanides, or rare earths, fits into Group 13. All are shiny metals.

This row, called the actinides, fits into Group 13. It includes radium and plutonium.

▲ There are eight Groups (columns) across in the Periodic Table and seven Periods (rows) down. The block of transition metals in the middle fits into Group 13. Each element has three electrons in its outer shell, but different numbers in the inner shells.

Crystals

- **Crystals** are particular kinds of solids that are made from a regular arrangement, or lattice, of atoms. Most rocks and metals are crystals, so are snowflakes and salt.

- **Most crystals** have regular, geometrical shapes with smooth faces and sharp corners.

- **Most crystals** grow in dense masses, as in metals. Some crystals grow separately, like grains of sugar.

- **Some crystals** are shiny and clear to look at. Crystals got their name from the chunks of quartz that the ancient Greeks called *krystallos*. They believed the chunks were unmeltable ice.

- **Crystals** form by a process called crystallization. As liquid evaporates or molten solids cool, the chemicals dissolved in them solidify.

▶ *Crystals such as these grow naturally as minerals are deposited from hot mineral-rich liquids underground.*

- **Crystals** grow as more atoms attach themselves to the lattice, just as icicles grow as water freezes onto them.

- **The smallest crystals** are microscopically small. Occasionally crystals of a mineral such as beryl may grow to the size of telegraph poles.

- **A liquid crystal** is a crystal that can flow like a liquid but has a regular pattern of atoms.

- **A liquid crystal** may change colour or go dark when the alignment of its atoms is disrupted by electricity or heat. Liquid crystal displays (LCDs) use a tiny electric current to make crystals affect light.

- **X-ray crystallography** uses X-rays to study the structure of atoms in a crystal. This is how we know the structure of many important life substances such as DNA.

Friction

- **Friction** is the force that acts between two things rubbing together. It stops them sliding past each other.

- **The friction** that stops things starting to slide is called static friction. The friction that slows down sliding surfaces is called dynamic friction.

- **The harder** two surfaces are pressed together, the greater the force that is needed to overcome friction.

- **The coefficient of friction (CF)** is the ratio of the friction to the weight of the sliding object.

- **Metal sliding on metal** has a CF of 0.74; ice sliding on ice has a CF of 0.1. This means it is over seven times harder to make metal slide on metal than ice on ice.

- **Friction often makes things hot**. As the sliding object is slowed down, much of the energy of its momentum is turned into heat.

- **Fluid friction** is the friction between moving fluids or between a fluid and a solid. It is what makes thick fluids viscous (less runny).

- **Oil reduces friction** by creating a film that keeps the solid surfaces apart.

- **Brakes use dynamic friction** to slow things down.

- **Drag is friction** between air and an object. It slows a fast car, or any aircraft moving through the air.

▼ *Waxed skis on snow have a CF of just 0.14, allowing cross-country skiers to slide along the ground very easily.*

Lavoisier

- **Antoine Laurent Lavoisier** (1743–1794) was a brilliant French scientist who is regarded as the founder of modern chemistry.

- **He was elected** to the French Royal Academy of Sciences at just 25 for an essay on street lighting. A year later, he worked on the first geological map of France.

- **Lavoisier earned his living** for a long while as a 'tax farmer', which meant he worked for a private company collecting taxes.

- **In 1771** he married 14-year-old Marie Paulze, who later became his illustrator and collaborator in the laboratory.

- **Lavoisier** was the first person to realize that air is essentially a mixture of two gases: oxygen and nitrogen.

- **Lavoisier discovered** that water is a compound of hydrogen and oxygen.

◀ *Lavoisier showed the importance of precision weighing in the laboratory.*

- **Lavoisier** showed that the popular phlogiston theory of burning was wrong and that burning involves oxygen instead.

- **Lavoisier** gave the first working list of chemical elements in his famous book *Elementary Treatise of Chemistry* (1789).

- **From 1776** Lavoisier headed research at the Royal Arsenal in Paris, developing gunpowder manufacture.

- **Lavoisier ran schemes** for public education, fair taxation, old-age insurance and other welfare schemes. But his good deeds did not save him. When Lavoisier had a wall built round Paris to reduce smuggling, revolutionary leader Marat accused him of imprisoning Paris's air. His past as a tax farmer was remembered and Lavoisier was guillotined in 1794.

Electromagnetism

▲ *Wind turbines generate electricity by using the wind to turn their blades. These drive magnets around inside coils of electric wire.*

- **Electromagnetism** is the combined effect of electricity and magnetism.

- **Every electric current** creates its own magnetic field.

- **Maxwell's screw rule** says that the magnetic field runs the same way a screw turns if you screw it in the direction of the electric current – clockwise.

- **An electromagnet** is a strong magnet that is only magnetic when an electric current passes through it. It is made by wrapping a coil of wire, called a solenoid, around a core of iron.

- **Electromagnets** are used in everything from ticket machines and telephones to loudspeakers.

- **Magnetic levitation** trains use strong electromagnets to carry the train on a cushion of magnetic repulsion.

- **When an electric wire** is moved across a magnetic field, an electric current is created, or induced, in the wire. This is the basis of every kind of electricity generation.

- **Fleming's right-hand rule** says that if you hold your right thumb, first and middle fingers at 90° to each other, your middle finger shows the direction of the induced current – if your thumb points in the direction the wire moves and your first finger points out the magnetic field.

- **Electromagnets** can be switched on and off, unlike permanent magnets.

- **Around every** electric or magnetic object is an area, or electromagnetic field, where its force is effective.

Heat

- **Heat is the energy** of moving molecules. The faster molecules move, the hotter it is.
- **When you hold your hand** over a heater the warmth is the assault of billions of fast-moving molecules of air.
- **Heat** is the combined energy of all the moving molecules; temperature is how fast they are moving.
- **The coldest temperature possible** is absolute zero, or −273.15°C, when molecules stop moving.
- **When you heat a substance** its temperature rises because heat makes its molecules move faster.
- **The same amount** of heat raises the temperature of different substances by different amounts.
- **Specific heat** is the energy needed, in joules, to heat a substance by 1°C.

> ★ STAR FACT ★
> Heat makes the molecules of a solid vibrate; it makes gas molecules zoom about.

- **Argon gas** gets hotter quicker than oxygen. The shape of oxygen molecules means they absorb some energy not by moving faster but by spinning faster.
- **Heat always spreads out** from its source. It heats up its surroundings while cooling down itself.

◀ *Fire changes the energy in fuel into heat energy. Heat makes the molecules rush about.*

Iron and steel

- **Iron** is the most common element in the world. It makes up 35% of the Earth, but most of it is in the Earth's core.
- **Iron is never found** in its pure form in the Earth's crust. Instead it is found in iron ores, which must be heated in a blast furnace to extract the iron.
- **The chemical symbol** for iron is Fe from *ferrum*, the Latin word for iron. Iron compounds are called either ferrous or ferric.
- **Iron has** an atomic number of 26 and an atomic weight of 55.85.

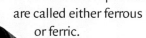
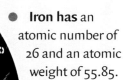

◀ *A solid-state laser can cut through carbon steel like butter even though steel is incredibly tough.*

- **Iron melts** at 1535°C and boils at 3000°C. It conducts heat and electricity quite well and dissolves in water very slowly. Iron is easily magnetized. It loses its magnetism easily but steel can be permanently magnetic.
- **Iron** combines easily with oxygen to form iron oxide, especially in the presence of moisture. This is rusting.
- **Cast iron** is iron with 2–4% carbon and 1–3% silicon. It is suitable for pouring into sand moulds. Wrought iron is almost pure iron with carbon removed to make it easy to bend and shape for railings and gates.
- **Iron is made into steel** by adding traces of carbon for making cars, railway lines, knives and much more. Alloy steels are made by adding traces of metals such as tungsten (for tools) and chromium (for ball bearings).
- **60% of steel** is made by the basic oxygen process in which oxygen is blasted over molten iron to burn out impurities.
- **Special alloy steels** such as chromium steels can be made from scrap iron (which is low in impurities) in an electric arc furnace.

Fibre optics

- **Fibre optic cables** are bundles of transparent glass threads that transmit messages by light.

- **The light is transmitted** in coded pulses.

- **A thin layer of glass,** called cladding, surrounds each fibre and stops light from escaping.

- **The cladding** reflects all the light back into the fibre so that it bends round with the fibre. This is called total internal reflection.

- **Single-mode fibres** are very narrow and the light bounces very little from side to side. These fibres are suitable for long-distance transmissions.

- **Aiming light** into the narrow core of a single-mode fibre needs the precision of a laser beam.

- **Multi-mode fibres** are wider than single-mode fibres. They accept LED (light-emitting diodes) light, so they are cheaper but they are unsuitable for long distances.

- **The largest cables** can carry hundreds of thousands of phone calls or hundreds of television channels.

▲ *A bundle of optical fibres glows with transmitted light.*

- **Underwater fibre optic** cables transmit signals under the Atlantic and Pacific Oceans.

- **Optical fibres** have medical uses, such as in endoscopes. These are flexible tubes with a lens on the end, which are inserted into the body to look inside it. Optical fibres are used to measure blood temperature and pressure.

Metals

▼ *Metals are very tough but can be easily shaped. They are used for an enormous variety of things, from chains to cars.*

- **75% of all known elements** are metals.

- **Most metals** ring when hit. A typical metal is hard but malleable, which means it can be hammered into sheets.

- **Metals** are usually shiny. They conduct both heat and electricity well.

- **Metals** do not form separate molecules. Instead atoms of metal knit together with metallic bonds (see chemical bonds) to form lattice structures.

- **The electron shells of all metals** are less than half-full. In a chemical reaction metals give up their electrons to a non-metal.

- **Most metals** occur naturally in the ground only in rocks called ores.

- **Gold, copper,** mercury, platinum, silver and a few other rare metals occur naturally in their pure form.

- **Mercury** is the only metal that is liquid at normal temperatures. It melts at −38.87°C.

- **A few atoms** of the new metal ununquadium (atomic number 114) were made in January 1999.

> ★ **STAR FACT** ★
> At 3410°C, tungsten has the highest melting point of any metal.

Radioactivity

- **Radioactivity** is when a certain kind of atom disintegrates spontaneously and sends out little bursts of radiation from its nucleus (centre).

- **Isotopes** are slightly different versions of an atom, with either more or less neutrons (see atoms). With stable elements, such as carbon, only certain isotopes called radio-isotopes are radioactive.

- **Some large atoms**, such as radium and uranium, are so unstable that all their isotopes are radio-isotopes.

- **Radioactive isotopes** emit three kinds of radiation: alpha, beta and gamma rays.

- **When the nucleus** of an atom emits alpha or beta rays it changes and becomes the atom of a different element. This is called radioactive decay.

- **Alpha rays** are streams of alpha particles. These are made from two protons and two neutrons – basically the nucleus of a helium atom. They travel only a few centimetres and can be stopped by a sheet of paper.

- **Beta rays** are beta particles. Beta particles are electrons (or their opposite, positrons) emitted as a neutron decays into a proton. They can travel up to 1 m and can penetrate aluminium foil.

▶ The radioisotope carbon-14 is present in all living things, but when they die it begins to steadily decay radioactively. By measuring how much carbon-14 is left relative to carbon-12 isotopes, which don't decay, scientists can tell exactly when a living thing died, including wood and bone.

- **Gamma rays** are an energetic, short-wave form of electromagnetic radiation (see electromagnetic spectrum). They penetrate most materials but lead.

- **The half-life** of a radioactive substance is the time it takes for its radioactivity to drop by half. This is much easier to assess than the time for the radioactivity to disappear altogether.

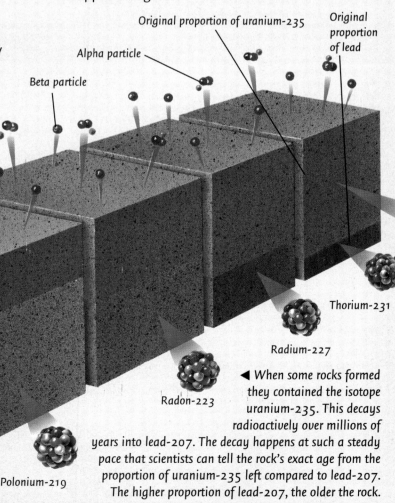

Original proportion of uranium-235

Original proportion of lead

Alpha particle

Beta particle

Thorium-231

Radium-227

Radon-223

Polonium-219

Lead-207

◀ When some rocks formed they contained the isotope uranium-235. This decays radioactively over millions of years into lead-207. The decay happens at such a steady pace that scientists can tell the rock's exact age from the proportion of uranium-235 left compared to lead-207. The higher proportion of lead-207, the older the rock.

Light and atoms

- **Light comes** from atoms. They give out light when they gain energy – by absorbing light or other electromagnetic waves or when hit by other particles.

- **Atoms** are normally in a 'ground' state. Their electrons circle close to the nucleus where their energy is at its lowest ebb.

- **An atom emits light** when 'excited' by taking in energy. Excitement boosts an electron's energy so it jumps further from the nucleus.

- **An atom** only stays excited a fraction of second before the electron drops back in towards the nucleus.

- **As an electron** drops back, it releases energy it gained as a packet of electromagnetic radiation called a photon.

- **Electrons** drop towards the nucleus in steps, like a ball bouncing down stairs.

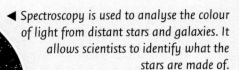

◀ *Spectroscopy is used to analyse the colour of light from distant stars and galaxies. It allows scientists to identify what the stars are made of.*

- **Since each step** the electron drops in has a particular energy level, so the energy of the photon depends precisely on how big the steps are. Big steps send out higher-energy short-wave photons like X-rays.

- **The colour of light** an atom sends out depends on the size of the steps its electrons jump down.

- **Each kind of atom** has its own range of electron energy steps, so each sends out particular colours of light. The range of colours each kind of atom sends out is called its emission spectrum. For gases, this acts like a colour signature to identify in a process called spectroscopy.

- **Just as an atom** only emits certain colours, so it can only absorb certain colours. This is its absorption spectrum.

Quantum physics

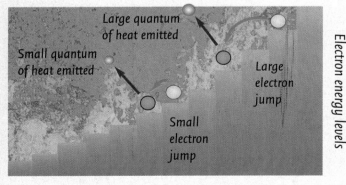

Large quantum of heat emitted

Small quantum of heat emitted

Large electron jump

Small electron jump

Electron energy levels

▲ *Quantum physics shows how radiation from a hot object is emitted in little chunks called quanta.*

- **By the 1890s** most scientists thought light moves in waves.

- **Max Planck** (1858–1947) realized that the range of radiation given out by a hot object is not quite what scientists would calculate it to be if radiation is waves.

- **Planck realized** that the radiation from a hot object could be explained if the radiation came in chunks, or quanta.

- **Quanta** are very, very small. When lots of quanta are

emitted together they appear to be like smooth waves.

- **In 1905** Einstein showed that quanta explain the photoelectric effect.

- **In 1913** Niels Bohr showed how the arrangement of electrons in energy levels around an atom (see electrons) could be thought of in a quantum way too.

- **In the 1920s** Erwin Schrödinger and Werner Heisenberg developed Bohr's idea into quantum physics, a new branch of physics for particles on the scale of atoms.

- **Quantum physics** explains how electrons emit radiation (see above). It shows that an electron is both a particle and a wave, depending on how you look at it. It seems to work for all four fundamental forces (see forces) except gravity.

- **The development** of the technologies that gave us lasers and transistors came from quantum physics.

- **Quantum physics** predicts some strange things on the scale of atoms, such as particles appearing from nowhere and electrons seeming to know where each other are.

Floating and sinking

- **Things float** because they are less dense than water, which is why you can lift quite a heavy person in a swimming pool. This loss of weight is called buoyancy.

- **Buoyancy** is created by the upward push, or upthrust, of the water.

- **When an object** is immersed in water, its weight pushes it down. At the same time the water pushes it back up with a force equal to the weight of water displaced (pushed out of the way). This is called Archimedes' principle (see Archimedes).

- **An object sinks** until its weight is exactly equal to the upthrust of the water, at which point it floats.

- **Things that are less dense** than water float; those that are more dense sink.

- **Steel ships** float because although steel is denser than water, their hulls are full of air. They sink until enough water is displaced to match the weight of steel and air in the hull.

- **Oil floats** on water because it is less dense.

- **Ships float** at different heights according to how heavily laden they are and how dense the water is.

- **Ships float higher** in sea water than in fresh water because salt makes the sea water more dense.

- **Ships float higher** in dense cold seas than in warm tropical ones. They float higher in the winter months.

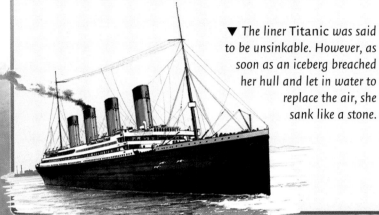

▼ The liner Titanic was said to be unsinkable. However, as soon as an iceberg breached her hull and let in water to replace the air, she sank like a stone.

Electronics

- **Electronics** are the basis of many modern technologies, from hi-fi systems to missile control systems.

- **Electronics** are systems that control things by automatically switching tiny electrical circuits on and off.

- **Transistors** are electronic switches. They are made of materials called semiconductors that change their ability to conduct electricity.

- **Electronic systems work** by linking many transistors together so that each controls the way the others work.

- **Diodes** are transistors with two connectors. They control an electric current by switching it on or off.

- **Triodes** are transistors with three connectors that amplify the electric current (make it bigger) or reduce it.

- **A silicon chip** is thousands of transistors linked by thin

▲ Microprocessors contain millions of transistors in a package that is no bigger than a human fingernail.

metal strips in an integrated circuit, on a single crystal of the semiconductor silicon.

- **The electronic areas** of a chip are those treated with traces of chemicals such as boron and phosphorus, which alter the conductivity of silicon.

- **Microprocessors** are complete Central Processing Units (see computers) on a single silicon chip.

> ★ STAR FACT ★
> Some microprocessors can now handle billions of bits of data every second.

Telecommunications

- **Telecommunications** is the almost instantaneous transmission of sounds, words, pictures, data and information by electronic means.

- **Every communication system** needs three things: a transmitter, a communications link and a receiver.

- **Transmitters** can be telephones or computers with modems (see the internet). They change the words, pictures, data or sounds into an electrical signal and send it. Similar receivers pick up the signal and change it back into the right form.

- **Communications links** carry the signal from the transmitter to the receiver in two main ways. Some give a direct link through telephone lines and other cables. Some are carried on radio waves through the air, via satellite or microwave links.

- **Telephone lines** used to be mainly electric cables which carried the signal as pulses of electricity. More and more are now fibre optics (see fibre optics) which carry the signal as coded pulses of light.

▼ *This illustration shows some of the many ways in which telecommunications are carried. At present, TV, radio and phone links are all carried separately, but increasingly they will all be carried the same way. They will be split up only when they arrive at their destination.*

- **Communications satellites** are satellites orbiting the Earth in space. Telephone calls are beamed up on radio waves to the satellite, which beams them back down to the right part of the world.

- **Microwave links** use very short radio waves to transmit telephone and other signals direct from one dish to another in a straight line across Earth's surface.

- **Mobile phones** or cellular phones transmit and receive phone calls directly via radio waves. The calls are picked up and sent on from a local aerial.

- **The information superhighway** is the network of high-speed links that might be achieved by combining telephone systems, cable TV and computer networks. TV programmes, films, data, direct video links and the Internet could all enter the home in this way.

> ★ STAR FACT ★
> Calls across the ocean go one way by satellite and the other by undersea cable to avoid delays.

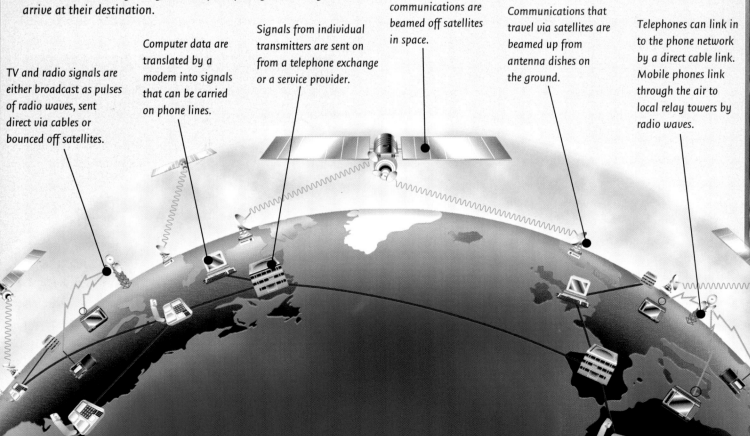

TV and radio signals are either broadcast as pulses of radio waves, sent direct via cables or bounced off satellites.

Computer data are translated by a modem into signals that can be carried on phone lines.

Signals from individual transmitters are sent on from a telephone exchange or a service provider.

More and more communications are beamed off satellites in space.

Communications that travel via satellites are beamed up from antenna dishes on the ground.

Telephones can link in to the phone network by a direct cable link. Mobile phones link through the air to local relay towers by radio waves.

Splitting the atom

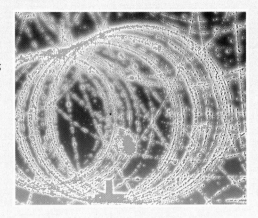

▶ You cannot actually see sub-atomic particles, but after collisions they leave tracks behind them, like these particle tracks in smoke. Photographs of these tracks tell scientists a great deal.

- **In the 1890s** scientists thought that atoms were solid like billiard balls and completely unbreakable.

- **In 1897** J. J. Thomson discovered that atoms contained even smaller particles, which he called electrons (see electrons).

- **In 1900** scientists thought atoms were like plum puddings with electrons like currants on the outside.

- **In 1909** Ernest Rutherford was firing alpha particles (see radioactivity) at a sheet of gold foil. Most went straight through, but 1 in 8000 particles bounced back!

- **Rutherford concluded** that the atom was mostly empty space (which the alpha particles passed straight through) but had a tiny, dense nucleus at its centre.

- **In 1919** Rutherford split the nucleus of a nitrogen atom with alpha particles. Small atoms could be split.

- **In 1932** James Chadwick found the nucleus contained two kinds of particle: protons and neutrons.

- **In 1933** Italian Enrico Fermi bombarded the big atoms of uranium with neutrons. Fermi thought the new atoms that formed had simply gained the neutrons.

- **In 1939** German scientists Hahn and Strassman repeated Fermi's experiment and found smaller atoms of barium.

- **Austrian Lise Meitner** realized that Hahn and Strassman had split uranium atoms. This discovery opened the way to releasing nuclear energy by fission.

Heat movement

- **Heat moves** in three different ways: conduction, convection and radiation.

- **Conduction** involves heat spreading from hot areas to cold areas by direct contact. It works a bit like a relay race. Energetic, rapidly moving or vibrating molecules cannon into their neighbours and set them moving.

- **Good conducting materials** such as metals feel cool to the touch because they carry heat away from your fingers quickly. The best conductors of heat are the metals silver, copper and gold, in that order.

- **Materials** that conduct heat slowly are called insulators. They help keep things warm by reducing heat loss. Wood is one of the best insulators. Water is also surprisingly effective as an insulator, which is why some divers and surfers often wear wetsuits.

- **Radiation** is the spread of heat as heat rays, that is, invisible waves of infrared radiation.

◀ Hot-air balloons work because hot air is lighter and rises through cold air.

- **Radiation** spreads heat without direct contact.

- **Convection** is when warm air rises through cool air, like a hot-air balloon.

- **Warm air rises** because warmth makes the air expand. As it expands the air becomes less dense and lighter than the cool air around it.

- **Convection currents** are circulation patterns set up as warm air (or a liquid) rises. Around the column of rising warmth, cool air continually sinks to replace it at the bottom. So the whole air turns over like a non-stop fountain.

> **★ STAR FACT ★**
> Convection currents in the air bring rain; convection currents in the Earth's interior move continents.

Light sources

▲ *Gas mixtures in neon lights glow different colours. Pure neon glows red.*

◄ *Electrical resistance makes the thin filament in a bulb glow.*

- **The light falling** on a surface is measured in lux. 1 lux is how brightly lit something is by a light of 1 candela 1 m away. You need 500 lux to read by.

- **Our main sources of natural light** are the Sun and the stars. The hot gases on their surface glow fiercely.

- **Electric lightbulbs** are incandescent, which means that their light comes from a thin tungsten wire, or filament, that glows when heated by an electric current.

- **The brightness** of a light source is measured in candelas (cd); 1 candela is about as bright as a small candle.

- **Lightbulbs** are filled with an inert (unreactive) gas, such as argon, to save the filament from burning out.

- **For 0.1 millisecond** an atom bomb flashes out 2000 billion candelas for every square metre (m²).

- **Electric lights** were invented independently in 1878 by Englishman Sir Joseph Swan and Americans Thomas Alva Edison and Hiram Maxim.

- **The Sun's surface** pumps out 23 billion candela per m². Laser lights are even brighter, but very small.

- **Fluorescent lights** have a glass tube coated on the inside with powders called phosphors. When electricity excites the gases in the tube to send out invisible UV rays, the rays hit the phosphors and make them glow, or fluoresce.

- **In neon lights,** a huge electric current makes the gas inside a tube electrically charged and so it glows.

Weight and mass

- **Mass** is the amount of matter in an object.

- **Weight** is not the same as mass. Scientists say weight is the force of gravity pulling on an object. Weight varies with the mass of the object and the strength of gravity.

- **Objects weigh more** at sea level, which is nearer the centre of the Earth, than up a mountain.

- **A person on the Moon** weighs one sixth of their weight on Earth because the Moon's gravity is one sixth of the Earth's gravity.

- **Weight varies** with gravity but mass is always the same, so scientists use mass when talking about how heavy something is.

► *Brass is used for weights as it is reasonably dense and and does not corrode.*

- **The smallest** known mass is that of a photon (see light and atoms). Its mass is 5.3 times 10^{-63} (62 zeros and a 1 after the decimal point) kg.

- **The mass of the Earth** is 6×10^{24} (six trillion trillion) kg. The mass of the Universe may be 10^{51} (10 followed by 50 zeros) kg.

- **Density is** the amount of mass in a certain space. It is measured in grams per cubic centimetre (g/cm³).

- **The lightest** solids are silica aerogels made for space science, with a density of 0.005 g/cm³. The lightest gas is hydrogen, at 0.00008989 g/cm³. The density of air is 0.00128 g/cm³.

- **The densest** solid is osmium at 22.59 g/cm³. Lead is 11.37 g/cm³. A neutron star has an incredible density of about one billion trillion g/cm³.

The Curies

- **Pierre and Marie Curie** were the husband and wife scientists who discovered the nature of radioactivity. In 1903 they won a Nobel Prize for their work .

- **Marie Curie** (1867–1934) was born Marya Sklodowska in Poland. She went to Paris in 1891 to study physics.

- **Pierre Curie** (1859–1906) was a French lecturer in physics who discovered the piezoelectric effect in crystals. His discovery led to the development of devices from quartz watches to microphones.

- **The Curies** met in 1894 while Marie was a student at the Sorbonne. They married in 1895.

- **In 1896** Antoine Becquerel found that uranium salts emitted a mysterious radiation that affected photographic paper in the same way as light.

- **In 1898** the Curies found the intensity of radiation was in exact proportion to the amount of uranium – so it must be coming from the uranium atoms.

- **The Curies called** atomic radiation 'radioactivity'.

- **In July 1898 t**he Curies discovered a new radioactive element. Marie called it polonium after her native Poland.

- **In December** the Curies found radium – an element even more radioactive than uranium.

- **In 1906** Pierre was killed by a tram. Marie died later from the effects of her exposure to radioactive materials, the dangers of which were unknown at that time.

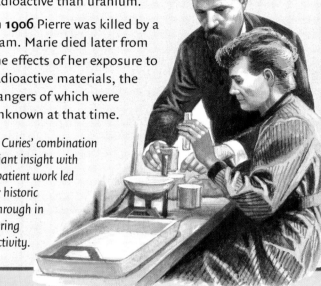

▶ *The Curies' combination of brilliant insight with exact, patient work led to their historic breakthrough in discovering radioactivity.*

Hydrogen

▲ *Hydrogen's combination with oxygen in water makes it one of the most important elements on the Earth.*

- **Hydrogen** is the lightest of all gases and elements. A large swimming pool of hydrogen would weigh just 1 kg.

- **Hydrogen** atoms have just one proton and one electron.

- **Hydrogen** is the first element in the Periodic Table. It has an Atomic Number of 1 and an atomic mass of 1.00794.

- **One in every 6000** hydrogen atoms has a neutron as

well as a proton in its nucleus, making it twice as heavy. This atom is called deuterium.

- **Rare** hydrogen atoms have two neutrons as well as the proton, making them three times as heavy. These are called tritium.

- **Hydrogen** is the most common substance in the Universe, making up over 90% of the Universe's weight.

- **Hydrogen** was the first element to form after the Universe began. It was billions of years before another element formed.

- **Most hydrogen** on Earth occurs in combination with other elements, such as oxygen in water. Pure hydrogen occurs naturally in a few places, such as small underground pockets and as tiny traces in the air.

- **Hydrogen** is one of the most reactive gases. It bursts easily and often explosively into flames.

- **Under extreme pressure** hydrogen becomes a metal – the most electrically conductive metal of all.

Quarks

- **Quarks** are one of the three tiniest basic, or elementary, particles from which every substance is made.

- **Quarks** are too small for their size to be measured, but their mass can. The biggest quark, called a top quark, is as heavy as an atom of gold. The smallest, called an up quark, is 35,000 times lighter.

- **There are six** kinds, or flavours, of quark: up (u), down (d), bottom (b), top (t), strange (s) and charm (c).

- **Down, bottom and strange** quarks carry one third of the negative charge of electrons; up, top and charm ones carry two-thirds of the positive charge of protons.

- **Quarks never exist** separately but in combination with one or two other quarks. Combinations of two or

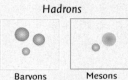

6 kinds of lepton   6 kinds of quark   3 kinds of bosons

▲ *These are the main kinds of elementary particle – quarks, leptons and bosons. Hadrons are combinations of two or three quarks.*

Hadrons

Baryons   Mesons

three quarks are called hadrons.

- **Three-quark hadrons** are called baryons and include protons and neutrons. Rare two-quark hadrons are mesons.

- **A proton** is made from two up quarks (two lots of +2/3 of a charge) and one down quark (−1/3) and has a positive charge of 1.

- **A neutron** is made from two down quarks (two lots of −1/3 of a charge) and an up quark (+2/3). The charges cancel each other out, giving a neutron no charge.

- **The theory of quarks** was first proposed by Murray Gell-Mann and Georg Zweig in 1964.

- **Quarks** are named after a famous passage in James Joyce's book *Ulysses*: 'Three quarks for Muster Mark!'

Air

- **The air** is a mixture of gases, dust and moisture.

- **The gas nitrogen** makes up 78.08% of the air. Nitrogen is largely unreactive, but it sometimes reacts with oxygen to form oxides of nitrogen.

- **Nitrogen** is continually recycled by the bacteria that consume plant and animal waste.

- **Oxygen** makes up 20.94% of the air. Animals breathe

> **★ STAR FACT ★**
> Air is a unique mixture that exists on Earth and nowhere else in the solar system.

in oxygen. Plants give it out as they take their energy from sunlight in photosynthesis.

- **Carbon dioxide** makes up 0.03% of the air. Carbon dioxide is continually recycled as it is breathed out by animals and taken in by plants in photosynthesis.

- **The air contains** other, inert (unreactive) gases: 0.93% is argon; 0.0018% is neon; 0.0005% is helium.

- **There are tiny traces** of krypton and xenon which are also inert.

- **Ozone makes up** 0.00006% of the air. It is created when sunlight breaks up oxygen.

- **Hydrogen makes up** 0.00005% of the air. This gas is continually drifting off into space.

Relativity

- **Einstein** (see Einstein) was the creator of two theories of relativity which have revolutionized scientists' way of thinking about the Universe: the special theory of relativity (1905) and the general theory (1915).

- **Time is relative** because it depends where you measure it from (see time). Distances and speed are relative too. If you are in a car and another car whizzes past you, for instance, the slower you are travelling, the faster the other car seems to be moving.

- **Einstein showed** in his special theory of relativity that you cannot even measure your speed relative to a beam of light, which is the fastest thing in the Universe. This is because light always passes you at the same speed, no matter where you are or how fast you are going.

> ★ **STAR FACT** ★
> When astronauts went to the Moon, their clock lost a few seconds. The clock was not faulty, but time actually ran slower in the speeding spacecraft.

- **Einstein** realized that if light always travels at the same speed, there are some strange effects when you are moving very fast (see below).

- **If a rocket** passing you zoomed up to near the speed of light, you would see it shrink.

- **If a rocket** passing you zoomed up to near the speed of light, you'd see the clocks on the rocket running more slowly as time stretched out. If the rocket reached the speed of light, the clocks would stop altogether.

- **If a rocket** accelerates towards the speed of light, its mass increases dramatically. This is because its momentum is its mass multiplied by its speed. As it still has the same momentum , it must be gaining mass.

- **Einstein's general relativity theory** brought in gravity. It showed that gravity works basically by bending space–time. From this theory scientists predicted black holes (see Hawking) and wormholes (see time travel).

- **In 1919** an eclipse of the Sun allowed Arthur Eddington to observe how the Sun bends light rays, proving Einstein's theory of General Relativity.

▼ In normal everyday life, the effects of relativity are so tiny that you can ignore them. However, in a spacecraft travelling very fast they may become quite significant.

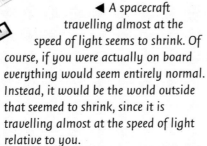

▲ In a spacecraft travelling almost at the speed of light, time runs slower. So astronauts going on a long, very fast journey into space come back a little younger than if they had stayed on the Earth.

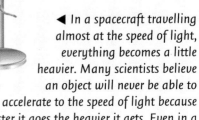

◄ A spacecraft travelling almost at the speed of light seems to shrink. Of course, if you were actually on board everything would seem entirely normal. Instead, it would be the world outside that seemed to shrink, since it is travelling almost at the speed of light relative to you.

◄ In a spacecraft travelling almost at the speed of light, everything becomes a little heavier. Many scientists believe an object will never be able to accelerate to the speed of light because the faster it goes the heavier it gets. Even in a fast-moving lift, things become very marginally heavier.

Hawking

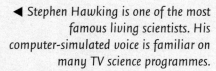

◀ *Stephen Hawking is one of the most famous living scientists. His computer-simulated voice is familiar on many TV science programmes.*

- **Stephen Hawking** (b.1942) is a British physicist who is famous for his ideas on space and time.

- **Hawking was born** in Oxford, England and studied at Cambridge University, where he is now a professor.

- **Hawking suffers** from the paralysing nerve disease called amyotrophic lateral sclerosis. He cannot move any more than a few hand and face muscles, but he gets around very well in an electric wheelchair.

- **Hawking cannot speak,** but he communicates effectively with a computer-simulated voice.

- **Hawking's** book *A Brief History of Time* (1988) outlines his ideas on space, time and the history of the Universe since the Big Bang. It was one of the best-selling science books of the 20th century.

- **Hawking's contributions** to the study of gravity are considered to be the most important since Einstein's.

- **More than anyone else,** Hawking has developed the idea of black holes – points in space where gravity becomes so extreme that it even sucks in light.

- **Hawking developed** the idea of a singularity, which is an incredibly small point in a black hole where all physical laws break down.

- **Hawking's work** provides a strong theoretical base for the idea that the Universe began with a Big Bang, starting with a singularity and exploding outwards.

- **Hawking** is trying to find a quantum theory of gravity (see quantum physics) to link in with the three other basic forces (electromagnetism and nuclear forces).

Oil

- **Oils** are liquids that do not dissolve in water and burn easily.

- **Oils are usually made** from chains of carbon and hydrogen atoms.

- **There are three main kinds of oil:** essential, fixed and mineral oils.

- **Essential oils** are thin, perfumed oils from plants. They are used in flavouring and aromatherapy.

▲ *Oil from underground and undersea sediments provides over half the world's energy needs.*

★ **STAR FACT** ★
Petroleum is used to make products from aspirins and toothpaste to CDs, as well as petrol.

- **Fixed oils** are made by plants and animals from fatty acids. They include fish oils and nut and seed oils.

- **Mineral oils** come from petroleum formed underground over millions of years from the remains of sea organisms.

- **Petroleum,** or crude oil, is made mainly of hydrocarbons – compounds of hydrogen and carbon, such as methane.

- **Hydrocarbons** in petroleum are mixed with oxygen, sulphur, nitrogen and other elements.

- **Petroleum** is separated by distillation into various substances such as aviation fuel, petrol or gasoline paraffin. As oil is heated in a distillation column, a mixture of gases evaporates. Each gas cools and condenses at different heights to a liquid, or fraction, which is then drawn off.

Scanners

- **Scanners** are electronic devices that move backwards and forwards in lines in order to build up a picture.

- **Image scanners** are used to convert pictures and other material into a digital form for computers to read.

- **A photoelectric cell** in the scanner measures the amount of light reflected from each part of the picture and converts it into a digital code.

- **Various complex scanners** are used in medicine to build up pictures of the inside of the body. They include CT scanners, PET scanners and MRI scanners.

- **CT stands** for computerized tomography. An X-ray beam rotates around the patient and is picked up by detectors on the far side to build up a 3-D picture.

- **PET** stands for Positron Emission Tomography. The scanner picks up positrons (positively charged electrons) sent out by radioactive substances injected into the blood.

- **PET scans** can show a living brain in action.

- **MRI** stands for Magnetic Resonance Imaging.

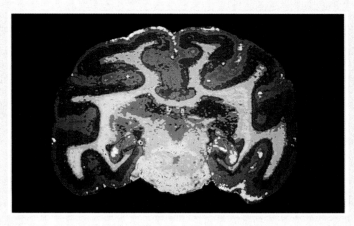

▲ *This PET scan shows a monkey's brain from above.*

- **An MRI scan** works like CT scans but it uses magnetism, not X-rays. The patient is surrounded by such powerful magnets that all the body's protons (see atoms) line up.

- **The MRI scan begins** as a radio pulse that knocks the protons briefly out of alignment. The scanner detects radio signals sent out by the protons as they snap back into line.

Vectors

- **For scientists,** vectors are things that have both a particular size and a particular direction.

- **Forces** such as gravity, muscles or the wind are vector quantities.

- **Acceleration** is a vector quantity.

- **Scalar quantities** are things which have a particular size but have no particular direction.

- **Speed, density and mass** are all scalar quantities.

- **Velocity** is speed in a particular direction, so it is a vector quantity.

- **A vector** can be drawn on a diagram with an arrow that is proportional in length to its size, pointing in the right direction.

- **Several vectors** may affect something at the same time. As you sit on a chair, gravity pulls you downwards while the chair pushes you up, so you stay still.

But if someone pushes the chair from behind, you may tip over. The combined effect of all the forces involved is called the resultant.

- **When several vectors** affect the same thing, they may act at different angles. You can work out their combined effect – the resultant – by drawing geometric diagrams with the vectors.

- **The parallelogram of forces** is a simple geometric diagram for working out the resultant from two forces. A vector arrow is drawn for each force from the same point. A parallel arrow is then drawn from the end of each arrow to make a parallelogram. The resultant is the simple diagonal of the parallelogram.

◀ *As a gymnast pauses in mid-routine, she unconsciously combines the forces acting on her to keep her in balance. These forces, such as her weight and forward momentum, are all vectors and could be worked out geometrically on paper.*

Genetic engineering

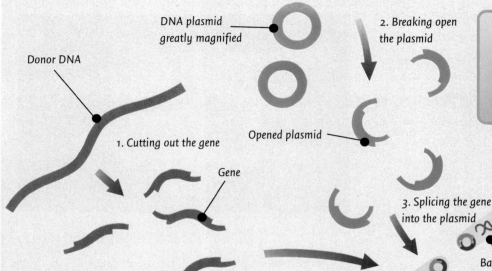

Donor DNA

DNA plasmid greatly magnified

1. Cutting out the gene

Gene

2. Breaking open the plasmid

Opened plasmid

3. Splicing the gene into the plasmid

Bacteria

Bacteria's ordinary DNA

Altered DNA plasmid less magnified

4. Bacteria with altered DNA multiplying

◀ This illustration shows the steps in gene splicing: 1. The bit of the donor DNA carrying the right gene is snipped out using restriction enzymes. 2. A special ring of DNA called a plasmid is then broken open. 3. The new gene is spliced into the plasmid, which is sealed up with DNA ligase and introduced into bacteria. 4. The bacteria then reproduce.

- **Genetic engineering** means deliberately altering the genes of plants and animals to give them slightly different life instructions.

- **Genes** are found in every living cell on special molecules called DNA (deoxyribonucleic acid). Engineering genes means changing the DNA.

- **Scientists alter genes** by snipping them from the DNA of one organism and inserting them into the DNA of another. This is called gene splicing. The altered DNA is called recombinant DNA.

- **Genes are cut** from DNA using biological scissors called restriction enzymes. They are spliced into DNA using biological glue called DNA ligase.

- **Once a cell** has its new DNA, every time the cell reproduces the new cells will have the same altered DNA.

- **By splicing new genes** into the DNA of bacteria, scientists can turn them into factories for making valuable natural chemicals. One protein made like this is interferon, a natural body chemical which protects humans against certain viruses. Enzymes for detergents and melanin for suntan lotion can be made like this too.

- **Scientists are now** finding ways of genetically modifying food crops. Crops may be engineered, for instance, to make them resistant to pests or frost. The first GM food was the 'Flavr Savr' tomato, which was introduced by the US biotechnology company Calgene in 1994.

- **Gene therapy** means altering the genes to cure diseases that are inherited from parents or caused by faulty genes.

- **Cloning** means creating an organism with exactly the same genes as another. Normally, new life grows from sex cells – cells from both parents – in which genes are mixed. Cloning takes DNA from any body cell and uses it to grow a new life. Since the new life has the same genes as the donor of the DNA, it is a perfect living replica.

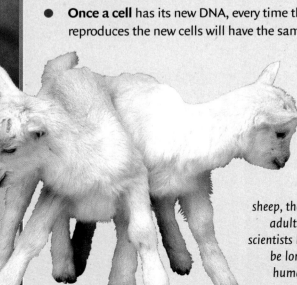

◀ In 1997, Ian Wilmut and his colleagues in Edinburgh, Scotland, created Dolly the sheep, the first clone of an adult mammal. Many scientists believe it will not be long before the first human clone is made.

Mixing colours

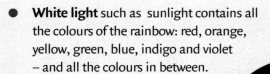

- **White light** such as sunlight contains all the colours of the rainbow: red, orange, yellow, green, blue, indigo and violet – and all the colours in between.

- **There are three basic,** or primary colours of light – red, green and blue. They can be mixed in various proportions to make any other colour.

- **The primary colours** of light are called additive primaries, because they are added together to make different colours.

- **Each additive primary** is one third of the full spectrum of white light, so adding all three together makes white.

- **When two additive primaries** are added together they make a third colour, called a subtractive primary.

- **The three subtractive primaries are:** magenta (red plus blue), cyan (blue plus green) and yellow (green plus red).

◀ These circles are the primary colours of light: red, green and blue. Where they overlap, you see the three subtractive primaries: magenta, cyan and yellow.

They too can be mixed in various proportions to make other colours.

- **Coloured surfaces** such as painted walls and pictures get their colour from the light falling on them. They soak up some colours of white light and reflect the rest. The colour you see is the colour reflected.

- **With reflected colours,** each subtractive primary soaks up one third of the spectrum of white light and reflects two-thirds of it. Mixing two subtractive primaries soaks up two-thirds of the spectrum. Mixing all three subtractive primaries soaks up all the spectrum, making black.

- **Two subtractive primaries** mixed make an additive primary.

- **Cyan and magenta** make blue; yellow and cyan make green; yellow and magenta make red.

Velocity and acceleration

- **Velocity** is speed in a particular direction.

- **Uniform velocity** is when the velocity stays the same. It can be worked out simply by dividing the distance travelled (d) by the time (t): $v = d/t$.

- **Acceleration** is a change in velocity.

- **Positive acceleration** is getting faster; negative acceleration, or deceleration, is getting slower.

- **Acceleration** is typically given in metres per second per second (m/s^2). In each second a velocity gets faster by so many metres per second per second.

- **A rifle bullet** accelerates down the barrel at 3000 m/s^2. A fast car accelerates at 6 m/s^2.

- **When an object falls,** the Earth's gravity makes it accelerate at 9.81 m/s^2. This is called g.

- **Acceleration** is often described in gs.

- **In a rocket taking off** at 1 g, the acceleration has little effect. But at 3 g, it is impossible to move your arms and legs; at 4.5 g you black out in five seconds.

- **A high-speed lift** goes up at 0.2 g. An plane takes off at 0.5 g. A car brakes at up to 0.7 g. In a crash, you may survive a momentary deceleration of up to 100 g, but the effects are likely to be severe.

▶ For a brief moment as they come away from the start, sprinters accelerate faster than a Ferrari car.

Organic chemistry

- **Organic chemistry** is the study of compounds that contain carbon atoms.

- **Over 90%** of all chemical compounds are organic.

- **Organic chemicals** are the basis of most life processes.

- **Scientists once thought** carbon compounds could only be made by living things. However, in 1828 Friedrich Wöhler made the compound urea in his laboratory.

- **By far the largest** group of carbon compounds are the hydrocarbons (see Oil).

- **Aliphatic organic compounds** are formed from long or branching chains of carbon atoms. They include ethane, propane and paraffin, and the alkenes from which many polymers are made (see oil compounds).

- **Cyclic organic compounds** are formed from closed rings of carbon atoms.

- **Aromatics** are made from a ring of six atoms (mostly carbon), with hydrogen atoms attached. They get their name from the strong aroma (smell) of benzene.

- **Benzene** is the most important aromatic. Friedrich Kekulé von Stradonitz discovered its six-carbon structure in 1865, after dreaming about a snake biting its own tail.

- **Isomers** are compounds with the same atoms but different properties. Butane and 2-methyl propane in bottled gas are isomers.

▼ *All living things are made basically of carbon compounds.*

Colour

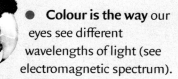

- **Colour is the way** our eyes see different wavelengths of light (see electromagnetic spectrum).

- **Red light** has the longest waves – about 700 nanometres, or nm (billionths of a metre).

- **Violet light** has the shortest waves – about 400 nm.

- **Light that is a mixture** of every colour, such as sunlight and the light from torches and ordinary lightbulbs, is called white light (see mixing colours).

◀ *The macaw gets its brilliant colours because pigment molecules in its feathers soak up certain wavelengths of light and reflect others, including reds, yellows and blues, very strongly.*

- **Things are different colours** because molecules in their surface reflect and absorb certain wavelengths of light.

- **Deep-blue printers inks** and bright-red blood are vividly coloured because both have molecules shaped like four-petalled flowers, with a metal atom at the centre.

- **Iridescence** is the shimmering rainbow colours you see flashing every now and then on a peacock's feathers, a fly's wings, oil on the water's surface or a CD.

- **Iridescence** can be caused by the way a surface breaks the light into colours like a prism does (see spectrum).

- **Iridescence** can be also caused by interference when an object has a thin, transparent surface layer. Light waves reflected from the top surface are slightly out of step with waves reflected from the inner surface, and they interfere.

★ STAR FACT ★
As a light source gets hotter, so its colour changes from red to yellow to white to blue.

Motion

- **Every movement** in the Universe seems to be governed by physical laws devised by people such as Isaac Newton and Albert Einstein.

- **Newton's first law of motion** says that an object accelerates, slows down or changes direction only when a force is applied.

- **Newton's second law of motion** says that the acceleration depends on how heavy the object is, and on how hard it is being pushed or pulled.

- **The greater the force** acting on an object, the more it will accelerate.

- **The heavier an object is** – the greater its mass – the less it will be accelerated by a particular force.

- **Newton's third law of motion** says that when a force pushes or acts one way, an equal force pushes in the opposite direction.

◄ To start moving, a skater uses the force of his muscles to push against the ground. As he or she pushes, the ground pushes back with equal force.

- **Newton's third law of motion** is summarized as follows: 'To every action, there is an equal and opposite reaction'.

- **Newton's third law** applies everywhere, but you can see it in effect in a rocket in space. In space there is nothing for the rocket to push on. The rocket is propelled by the action and reaction between the hot gases pushed out by its engine and the rocket itself.

- **You cannot always see** the reaction. When you bounce a ball on the ground, it looks as if only the ball bounces, but the Earth recoils too. The Earth's mass is so great compared to the ball's that the recoil is tiny.

- **Einstein's theory** of relativity modifies Newton's second law of motion under certain circumstances.

Molecules

- **A molecule** is two or more atoms bonded together. It is normally the smallest bit of a substance that exists independently.

- **Hydrogen atoms** exist only in pairs, or joined with atoms of other elements. A linked pair of hydrogen atoms is a hydrogen molecule.

- **Molecule atoms** are held together by chemical bonds.

- **The shape of a molecule** depends on the arrangement of

★ STAR FACT ★
If the DNA molecule in every human body cell were as thick as a hair, it would be 8 km long.

bonds that hold its atoms together.

- **Ammonia molecules** are pyramid shaped; some protein molecules are long spirals.

- **Compounds** only exist as molecules. If the atoms in the molecule of a compound were separated, the compound would cease to exist.

- **Chemical formulas** show the atoms in a molecule.

- **The formula for ammonia,** a choking gas, is NH_3, because an ammonia molecule is made from one nitrogen atom and three hydrogen atoms.

- **The mass of a molecule** is called the molecular mass. It is worked out by adding the mass of all the atoms in it.

◄ A crystal such as this is built from billions of identical molecules.

Space

- **A flat or plane surface** has just two dimensions at right angles to each other: length and width.

- **Any point** on a flat surface can be pinpointed exactly with just two figures: one showing how far along it is and the other how far across.

- **There are three** dimensions of space at right angles to each other: length, width and height.

- **A box** can be described completely with just one figure for each dimension.

- **A point in space** can be pinpointed with three figures: one shows how far along it is, one how far

▶ *An eclipse of the Sun in 1919 showed Einstein's suggestion that gravity can bend light is true. In bending light, gravity is also bending space–time.*

- across it is and a third how high up or down it is.

- **If something is moving**, three dimensions are not enough to locate it. You need a fourth dimension – time – to describe where the object is at a particular time.

- **In the early 1900s,** mathematician Hermann Minkowski realized that for Einstein's relativity theory you had to think of the Universe in terms of four dimensions, including time.

- **Four-dimensional** space is now called space–time.

- **Einstein's** theory of general relativity shows that space–time is actually curved.

- **After Minkowski's ideas,** mathematicians began to develop special geometry to describe four or even more dimensions.

Elements

- **Elements** are the basic chemicals of the Universe. Each one is made from one kind of atom, with a certain number of sub-atomic particles and its own unique character.

- **More than 115** elements have so far been identified.

- **Each element** is listed in the periodic table.

- **At least 20** of the most recently identified elements were created entirely by scientists and do not exist naturally.

> **! NEWS FLASH !**
> Scientists have made three atoms of a new element, 118 or ununoctium, which is probably a colourless gas.

- **All the most recently discovered elements** have very large, heavy atoms.

- **The lightest atom** is hydrogen.

- **The heaviest naturally occuring** element is osmium.

- **When different elements combine** they make chemical compounds (see chemical compounds).

- **New elements** get temporary names from their atomic number (see the Periodic Table). So the new element with atomic number 116 is called ununhexium. *Un* is the Latin word for one; *hex* is Latin for six.

◀ *Very few elements occur naturally by themselves. Most occur in combination with others in compounds. Gold is one of the few elements found as a pure 'native' element.*

Light

▶ This straw is not a light source, so we see it by reflected light. As the light rays reflected from the straw leave the water, they are bent, or refracted, as they emerge from the water and speed up. So the straw looks broken even though it remains intact.

- **Light is a form of energy.** It is one of the forms of energy sent out by atoms when they become excited.

- **Light is just one** of the forms of electromagnetic radiation (see electromagnetic spectrum). It is the only form that we can see.

- **Although we are surrounded** by light during the day, very few things give out light. The Sun and other stars and electric lights are light sources, but we see most things only because they reflect light. If something does not send out or reflect light, we cannot see it.

- **Light beams** are made of billions of tiny packets of energy called photons (see moving light). Together, these photons behave like waves on a pond. But the waves are tiny – 2000 would fit across a pinhead.

- **Light travels** in straight lines. The direction can be changed when light bounces off something or passes through it, but it is always straight. The straight path of light is called a ray.

- **When the path of a light ray** is blocked altogether, it forms a shadow. Most shadows have two regions: the umbra and penumbra. The umbra is the dark part where light rays are blocked altogether. The penumbra is the lighter rim where some rays reach.

- **When light rays** hit something, they bounce off, are soaked up or pass through. Anything that lets light through, such as glass, is transparent. If it mixes the light on the way through, such as frosted glass, it is translucent. If it stops light altogether, it is opaque.

- **When light strikes a surface**, some or all of it is reflected. Most surfaces scatter light in all directions, and all you see is the surface. But mirrors and other shiny surfaces reflect light in exactly the same pattern in which it arrived, so you see a mirror image.

- **When light passes** into transparent things such as water, rays are bent, or refracted. This happens because light travels more slowly in glass or water, and the rays swing round like the wheels of a car running onto sand.

▶ Glass lenses are shaped to refract light rays in particular ways. Concave lenses are dish-shaped lenses – thin in the middle and fat at the edges. As light rays pass through a concave lens they are bent outwards, so they spread out. The result is that when you see an object through a concave lens, it looks smaller than it really is.

▶ Convex lenses bulge outwards. They are fatter in the middle and thin around the edges. As light rays pass through a convex lens they are bent inwards, so they come together, or converge. When you see an object through a convex lens, it looks magnified. The point where the converging light rays meet is called the focus.

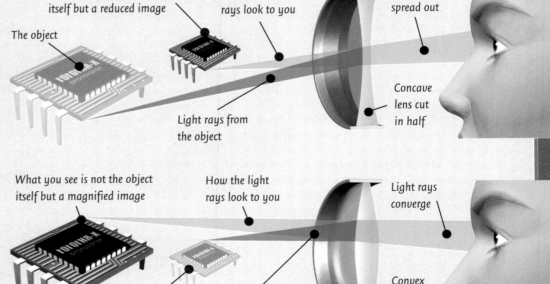

What you see is not the object itself but a reduced image

The object

How the light rays look to you

Light rays spread out

Light rays from the object

Concave lens cut in half

What you see is not the object itself but a magnified image

How the light rays look to you

Light rays converge

The object

Light rays from the object

Convex lens cut in half

Television

▲ *TV cameras convert a scene into an electrical signal.*

- **Television relies** on the photoelectric effect – the emission of electrons by a substance when struck by photons of light. Light-sensitive photocells in cameras work like this.

- **TV cameras** have three sets of tubes with photocells (reacting to red, green and blue light) to convert the picture into electrical signals.

- **The sound signal** from microphones is added, and a 'sync pulse' is put in to keep both kinds of signal in time.

- **The combined signal** is turned into radio waves (see electromagnetic spectrum) and broadcast.

- **An aerial** picks up the signal and feeds it to your television set.

- **Most TV sets** are based on glass tubes shaped like giant lightbulbs, called cathode-ray tubes. The narrow end contains a cathode, which is a negative electrical terminal. The wide end is the TV screen.

- **The cathode** fires a non-stop stream of electrons (see electrons) at the inside of the TV screen.

- **Wherever electrons** hit the screen, the screen glows as its coating of phosphors heats up.

- **To build up the picture** the electron beam scans quickly back and forth across the screen, making it glow in certain places. This happens so quickly that it looks as if the whole screen is glowing.

- **Colour TVs** have three electron guns: one to make red phosphors glow, another for green and a third for blue.

Huygens

- **Christiaan Huygens** (1629–1695) was, after Isaac Newton, the greatest scientist of the 1600s.

- **Huygens** was born to a wealthy Dutch family in The Hague, in Holland.

- **He studied law** at the University of Leiden and the College of Orange in Breda before turning to science.

- **He worked** with his brother Constanijn to grind lenses for very powerful telescopes.

- **With his powerful telescope,** Huygens discovered in 1655 that what astronomers had thought were Saturn's 'arms' were actually rings. He made his discovery known to people in code.

- **Huygens discovered** Titan, one of Saturn's moons.

◄ *Christiaan Huygens was the leading figure of the Golden Age of Dutch science in the 17th century, making contributions in many fields.*

- **Huygens** built the first accurate pendulum clock.

- **Responding to Newton's theory** that light was 'corpuscles', Huygens developed the theory that light is waves (see moving light) in 1678.

- **Huygens** described light as vibrations spreading through a material called ether, which is literally everywhere and is made of tiny particles. The idea of ether was finally abandoned in the late 19th century, but not the idea of light waves.

- **Huygens' wave idea** enabled him to explain refraction (see light) simply. It also enabled him to predict correctly that light would travel more slowly in glass than in air.

Engines

- **Engines are devices** that convert fuel into movement.

- **Most engines** work by burning the fuel to make gases that expand rapidly as they get hot.

- **Engines** that burn fuel to generate power are called heat engines. The burning is called combustion.

- **An internal combustion** engine, as used in a car, a jet or a rocket, burns its fuel on the inside.

▶ The Thrust car used a jet engine to give it the acceleration it needed for its attempt on the world land speed record.

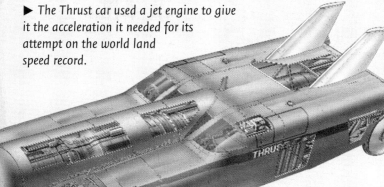

- **In engines** such as those in cars and diesel trains, the hot gases swell inside a chamber (the combustion chamber) and push against a piston or turbine.

- **An external combustion** engine burns its fuel on the outside in a separate boiler that makes hot steam to drive a piston or turbine. Steam engines on trains and boats work in this way.

- **Engines with pistons** that go back and forth inside cylinders are called reciprocating engines.

- **In jets and rockets,** hot gases swell and push against the engine as they shoot out of the back.

- **In four-stroke engines,** such as those in most cars, the pistons go up and down four times for each time they are thrust down by the hot gases.

- **In two-stroke engines,** such as those on small motorcycles and lawnmowers, the piston is pushed by burning gases every time it goes down.

Solutions

- **Tap water** is rarely pure water; it usually contains invisible traces of other substances. This makes it a solution.

- **A solution** is a liquid that has a solid dissolved within it.

- **When a solid dissolves,** its molecules separate and mix completely with the molecules of the liquid.

- **The liquid** in a solution is called the solvent.

- **The solid** dissolved in a solution is the solute.

- **The more of a solid that dissolves,** the stronger the solution becomes until at last it is saturated and no more will dissolve. There is literally no more room in the liquid.

- **If a saturated** solution is heated the liquid expands, making room for more solute to dissolve.

> ★ STAR FACT ★
> Ancient alchemists searched for a universal solvent in which all substances would dissolve.

▶ There are at least 30 common solutes in sea water. They include simple salt (sodium chloride) and magnesium chloride, but most are in very tiny amounts.

- **If a saturated** solution cools or is left to evaporate there is less room for solute, so the solute is precipitated (comes out of the solution).

- **Precipitated solute** molecules often link together to form solid crystals.

Nuclear power

- **Nuclear power** is based on the huge amounts of energy that bind together the nucleus of every atom in the Universe. It is an incredibly concentrated form of energy.

- **Nuclear energy** is released by splitting the nuclei of atoms in a process called nuclear fission (see nuclear energy). One day scientists hope to release energy by nuclear fusion – by fusing nuclei together as in the Sun.

- **Most nuclear reactors** use uranium-235. These are special atoms, or isotopes, of uranium with 235 protons and neutrons in their nucleus rather than the normal 238.

- **The fuel** usually consists of tiny pellets of uranium dioxide in thin tubes, separated by sheets called spacers.

- **Three kilograms of uranium fuel** provide enough energy for a city of one million people for one day.

▼ Like coal- and oil-fired power stations, nuclear power stations use steam to drive turbines to generate electricity. The difference is that nuclear power stations obtain the heat by splitting uranium atoms, not by burning coal or oil. When the atom is split, it sends out gamma rays, neutrons and immense heat. In a nuclear bomb this happens in a split second. In a nuclear power plant, control rods soak up some of the neutrons and slow the process down.

- **The earliest reactors,** called N-reactors, were designed to make plutonium for bombs. Magnox reactors make both plutonium and electricity.

- **Pressurized water reactors** (PWRs), originally used in submarines, are now the most common kind. They are built in factories, unlike Advanced Gas Reactors (AGRs).

- **Fast-breeder reactors** actually create more fuel than they burn, but the new fuel is highly radioactive.

- **Every stage of the nuclear process** creates dangerous radioactive waste. The radioactivity may take 80,000 years to fade. All but the most radioactive liquid waste is pumped out to sea. Gaseous waste is vented into the air. Solid waste is mostly stockpiled underground. Scientists debate fiercely about what to do with radioactive waste.

> ★ STAR FACT ★
> One kilogram of deuterium (a kind of hydrogen) can give the same amount of energy as three million kilograms of coal.

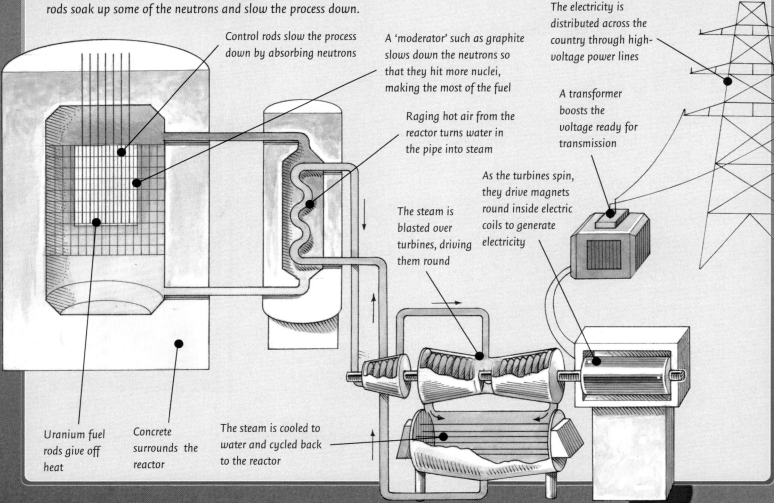

Control rods slow the process down by absorbing neutrons

A 'moderator' such as graphite slows down the neutrons so that they hit more nuclei, making the most of the fuel

The electricity is distributed across the country through high-voltage power lines

Raging hot air from the reactor turns water in the pipe into steam

A transformer boosts the voltage ready for transmission

As the turbines spin, they drive magnets round inside electric coils to generate electricity

The steam is blasted over turbines, driving them round

Uranium fuel rods give off heat

Concrete surrounds the reactor

The steam is cooled to water and cycled back to the reactor

Moving particles

- **The atoms and molecules** from which every substance is made are always moving.
- **The speed** at which they move depends on temperature.
- **Heat gives atoms and molecules** extra energy, making them move faster.
- **In 1827** Scottish botanist Robert Brown saw through a microscope that pollen grains in water were constantly dancing. They are buffeted by moving molecules that are too small to be seen. The effect is called Brownian motion.
- **In a gas,** the atoms and molecules are so far apart that they zoom about freely in all directions.
- **Smells spread** quickly because the smell molecules move about very quickly.

◀ As liquids boil, atoms and molecules move more and more energetically until some break away and turn to gas. This is evaporation.

- **In a liquid,** molecules are closely packed and move like dancers in a nightclub. If molecules stopped moving in liquids we would all die, because this movement is what moves materials in and out of human cells.
- **In a solid**, atoms and molecules are bound together and vibrate on the spot.
- **Air and water pressure** is simply bombardment by billions of moving molecules.
- **At −273.15°C,** absolute zero, the movement of atoms and molecules slows down to a complete standstill.

Inertia and momentum

- **Everything that is standing still** has inertia, which means that it will not move unless forced to.
- **Everything that moves** has momentum – it will not slow down, speed up or change direction unless forced to.
- **There is no real difference** between inertia and momentum, because everything in the universe is moving. Things only appear to be still because they are not moving relative to something else.
- **An object's momentum** is its mass times its velocity.
- **Something heavy** or fast has a lot of momentum, so a large force is needed to slow it down or speed it up.
- **A ball moves** when you kick it because when a moving object strikes another, its momentum is transferred. This is the law of conservation of momentum.
- **Angular momentum** is the momentum of something spinning. It depends on its speed and the size of the circle.
- **When a spinning skater** draws his arms close to his body, the circle he is making is smaller yet his angular

momentum must be the same. So he spins faster.
- **For the same reason a satellite** orbiting close to the Earth travels faster than one orbiting farther out.
- **A spinning top stays** upright because its angular momentum is greater than the pull of gravity.

▼ The lead shot that athletes throw when they put the shot has a large mass. It takes a lot of muscle power to overcome its inertia.

Holograms

- **Holograms** are three-dimensional photographic images made with laser lights.

- **The idea of holograms** was suggested by Hungarian-born British physicist Dennis Gabor in 1947. The idea could not be tried until laser light became available.

- **The first holograms** were made by Emmett Leith and Juris Upatnieks in Michigan, USA in 1963 and by Yuri Denisyuk in the Soviet Union.

- **To make a hologram,** the beam from a laser light is split in two. One part of the beam is reflected off the subject onto a photographic plate. The other, called the reference beam, shines directly onto the plate.

- **The interference** between light waves in the reflected beam and light waves in the reference beam creates the hologram in complex microscopic stripes on the plate.

- **Some holograms** only show up when laser light is shone through them.

- **Some holograms** work in ordinary light, such as those

▲ Holograms seem to hover strangely in space.

used in credit cards to stop counterfeiting.

- **Holograms** are used to detect defects in engines and aeroplanes, and forgeries in paintings by comparing two holograms made under slightly different conditions.

- **Huge amounts of digital data** can be stored in holograms in a crystal.

- **In 1993** 10,000 pages of data were stored in a lithium nobate crystal measuring just 1 cm across.

Sound measurement

▲ Heavy traffic is about 90 decibels, but it can rise higher.

- **The loudness of a sound** is usually measured in units called decibels (dB).

- **One decibel** is one tenth of a bel, the unit of sound named after Scots-born inventor Alexander Graham Bell.

- **Decibels** were originally used to measure sound intensity. Now they are used to compare electronic power output and voltages too.

- **Every 10 points up** on the decibel scale means that a sound has increased by ten times.

- **1 dB** is the smallest change the human ear can hear.

- **The quietest sound** that people can hear is 10 dB.

- **Quiet sounds:** a rustle of leaves or a quiet whisper is 10 dB. A quiet conversation is 30–40 dB. Loud conversation is about 60 dB. A city street is about 70 dB.

- **Loud sounds:** thunder is about 100 dB. The loudest scream ever recorded was 128.4 dB. A jet taking off is 110–140 dB. The loudest sound ever made by human technology (an atom bomb) was 210 dB.

- **The amount of energy** in a sound is measured in watts per square metre (W/m²). Zero dB is one thousand billionths of 1 W/m².

> ★ STAR FACT ★
> Sounds over 130 dB are painful; sounds over 90–100 dB for long periods cause deafness.

Electric power

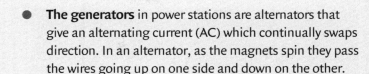

- **Most electricity is generated in power stations** by magnets that spin between coils of wire to induce an electric current (see electric circuits).

- **The magnets** are turned by turbines, which are either turned by steam heated by burning coal, oil or gas, or by nuclear fuel, or turned by wind or water.

- **The stronger the magnet**, the faster it turns and the more coils there are, so the bigger the voltage created.

- **Simple dynamos** generate a direct current (DC) that always flows in the same direction.

- **The generators** in power stations are alternators that give an alternating current (AC) which continually swaps direction. In an alternator, as the magnets spin they pass the wires going up on one side and down on the other.

- **The system of power transmission** that takes electricity into homes was developed by Croatian-born US engineer Nikola Tesla at Niagara, USA in the 1880s.

- **Electricity from power stations** is distributed around a country in a network of cables known as the grid.

- **Power station** generators push out 25,000 volts or more. This voltage is too much to use in people's homes, but not enough to transmit over long distances.

- **To transmit** electricity over long distances, the voltage is boosted to 400,000 volts by step-up transformers. It is fed through high-voltage cables. Near its destination the electricity's voltage is reduced by step-down transformers at substations for distribution to homes, shops, offices and factories.

▼ Electricity is brought to our homes through a network of high-tension cables. Some cables are buried underground, some are suspended high in the air from metal towers called pylons.

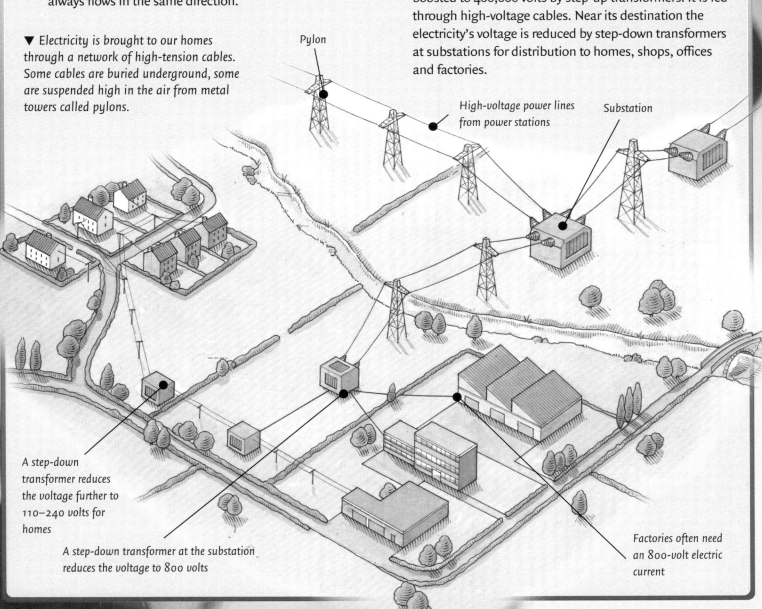

Pylon

High-voltage power lines from power stations

Substation

A step-down transformer reduces the voltage further to 110–240 volts for homes

A step-down transformer at the substation reduces the voltage to 800 volts

Factories often need an 800-volt electric current

Copper

▲ The high conductivity of copper makes it a perfect material for the core of electrical cables.

- **Copper** was one of the first metals used by humans over 10,000 years ago.
- **Copper** is one of the few metals that occur naturally in a pure form.
- **Most of the copper** that we use today comes from ores such as cuprite and chalcopyrite.

- **The world's biggest deposits** of pure copper are in volcanic lavas in the Andes Mountains in Chile.
- **Copper has** the atomic number 29, an atomic mass of 63.546 and melts at 1083°C.
- **Copper is** by far the best low-cost conductor of electricity, so it is widely used for electrical cables.
- **Copper is also** a good conductor of heat, which is why it is used to make the bases of saucepans.
- **Copper is so ductile** (easily stretched) that a copper rod as thick as a finger can be stretched out thinner than a human hair.
- **After being in the air** for some time, copper gets a thin green coating of copper carbonate called verdigris. 'Verdigris' means green of Greece.

★ STAR FACT ★
Copper is mixed with tin to make bronze, the an alloys dating back more than 5000 years.

Energy

- **Energy is the ability** to make things happen or, as scientists say, do work.
- **Energy comes in many forms,** from the chemical energy locked in sugar to the mechanical energy in a speeding train.
- **Energy does its work** either by transfer or conversion.

▲ Power stations do not create energy. They simply convert it into a convenient form for us to use – electricity.

- **Energy transfer** is the movement of energy from one place to another, such as heat rising above a fire or a ball being thrown.
- **Energy conversion** is when energy changes from one form to another – as when wind turbines generate electric power, for instance.
- **Energy is never lost nor gained;** it simply moves or changes. The total amount of energy in the Universe has stayed the same since the dawn of time.
- **Energy and mass** are actually the same thing. They are like flip sides of a coin and are interchangeable.
- **Potential energy** is energy stored up ready for action – as in a squeezed spring or a stretched piece of elastic.
- **Kinetic energy** is energy that something has because it is moving, such as a rolling ball or a falling stone.
- **Kinetic energy** increases in proportion with the velocity of an object squared. So a car has four times more kinetic energy at 40 km/h than at 20 km/h.

Thermodynamics

- **Energy cannot be destroyed** but it can be burned up. Every time energy is used, some turns into heat. This is why you feel hot after running.

- **Energy that turns into heat** may not be lost. It dissipates (spreads out thinly in all directions) and is hard to use again.

◀ *Waste gases produced from this chemial plant are burnt off and released into the atmosphere as unwanted heat.*

- **Every time energy is used,** some free energy (energy available to make things happen) gets lost as heat.

- **Scientists use** the word 'entropy' to describe how much energy has become unusable. The less free energy there is, the greater the entropy.

- **The word 'entropy'** was invented by the German physicist Rudolf Clausius in 1865.

- **Clausius showed** that everything really happens because energy moves from areas of high energy to areas of low energy, from hot areas to cold areas.

- **Energy goes on flowing** from high to low until there is no difference to make anything happen. This is an equilibrium state. Entropy is the maximum.

- **Clausius summed this idea up** in the 1860s with two laws of thermodynamics.

- **The first law of thermodynamics** says the total energy in the Universe was fixed forever at the beginning of time.

- **The second law of thermodynamics** says that heat cannot by itself pass from a colder object or place to a warmer one.

Sound

- **Most sounds** you hear, from the whisper of the wind to the roar of a jet, are simply moving air. When any sound is made it makes the air vibrate, and these vibrations carry the sound to your ears.

- **The vibrations** that carry sound through the air are called sound waves.

- **Sound waves** move by alternately squeezing air molecules together and then stretching them apart.

- **The parts of the air** that are squeezed are called condensations; the parts of the air that are stretched are called rarefactions.

- **Sound waves** travel faster through liquids and solids than through air because the molecules are more closely packed together in liquids and solids.

- **In a vacuum** such as space there is complete silence because there are no molecules to carry the sound.

- **Sound travels** at about 344 m per second in air at 20°C.

- **Sound travels** faster in warm air, reaching 386 metres per second at 100°C.

- **Sound travels** at 1500 metres per second in water and about 6000 metres per second in steel.

- **Sound travels a million times** slower than light, which is why you hear thunder after you see a flash of lightning, even though they both happen at the same time.

▶ *When you sing, talk or shout, you are actually vibrating the vocal cords in your throat. These set up sound waves in the air you push up from your lungs.*

Time travel

- **Einstein showed** that time runs at different speeds in different places – and is just another dimension. Ever since, some scientists have wondered whether we could travel through time to the past or the future (see time).

- **Einstein said** you cannot move through time because you would have to travel faster than light. If you travelled as fast as light, time would stop and you would not exist.

- **A famous argument** against time travel is about killing your grandparents. What if you travelled back in time before your parents were born and killed your grandparents? Then neither your parents nor you could have been born. So who killed your grandparents?

- **In the 1930s** American mathematician Kurt Gödel found time travel might be possible by bending space–time.

- **Scientists have come up** with all kinds of weird ideas for bending space–time, including amazing gravity machines. The most powerful benders of space–time are black holes in space.

- **Stephen Hawking** said you cannot use black holes for time travel because everything that goes into a black hole shrinks to a singularity (see Hawking). Others say you might dodge the singularity and emerge safely somewhere else in the Universe through a reverse black hole called a white hole.

- **US astronomer Carl Sagan** thought small black hole–white hole tunnels might exist without a singularity. There might be tunnels such as these linking different parts of the Universe, like a wormhole in an apple.

- **The mathematics** says that a wormhole would snap shut as soon as you stepped into it. However, it might be possible to hold it open with an anti-gravity machine based on a quantum effect called the Casimir effect.

- **Stephen Hawking** says wormholes are so unstable that they would break up before you could use them to time travel. Martin Visser says you might use them for faster than light (FTL) travel, but not for time travel.

▼ If wormholes exist, they are thought to be very, very tiny – smaller than an atom. So how could they be used for time travel? Some scientists think you may be able to use an incredibly powerful electric field to enlarge them and hold them open long enough to make a tunnel through space–time.

2000

The wormhole time machine depends on blowing a wormhole up large enough and holding it open long enough for you to slip through.

Although it's hard to imagine, space–time is not space with stars like this at all. It is a four-dimensional space and travelling through space–time means travelling through time as well as space.

The far end of a wormhole is the opposite of a black hole – a white hole. It pushes things out, not sucks them in.

1500

If you could create a wormhole time machine, just where and when would the other end be?

! NEWS FLASH !
A new theory called the Many Worlds Theory suggests that at any moment there are many different possible futures – and all of these futures actually happen.

Sound recording

▲ In a recording studio, the sound is recorded either on computer or on big master tapes.

- **Sound is recorded** by using a microphone to turn the vibrations of sound into a varying electrical current.

- **Sound recording** in the past was analogue, which means that the electrical current varies continually exactly as the sound vibrations do.

- **Most sound recording** today is digital, which means that sound vibrations are broken into tiny electrical chunks.

- **To make a digital recording** a device called an analogue-to-digital converter divides the vibrations into 44,100 segments for each second of sound.

- **Each digital segment** is turned into a digital code of ON or OFF electrical pulses.

- **With analogue sound,** each time the signal is passed on to the next stage, distortion and noise are added. With digital sound no noise is added, but the original recording is not a perfect replica of the sound.

- **On a CD (compact disc)** the pattern of electrical pulses is burned by laser as a pattern of pits on the disc surface.

- **During playback,** a laser beam is reflected from the tiny pits on a CD to re-create the electrical signal.

- **DVDs** work like CDs. They can store huge amounts of data on both sides, but most can only be recorded on once.

- **Minidiscs** (MDs) use magneto-optical recording to allow you to record on the disc up to one million times. A laser burns the data into a magnetic pattern on the disc.

Glass

- **Glass** is made mainly from ordinary sand (made of silica), from soda ash (sodium carbonate) and from limestone (calcium carbonate).

- **Glass** can be made from silica alone. However, silica has a very high melting point (1700°C), so soda ash is added to lower its melting point.

- **Adding a lot of soda ash** makes glass too soluble in water, so limestone is added to reduce its solubility.

- **To make sheets of glass,** 6% lime and 4% magnesia (magnesium oxide) are added to the basic mix.

- **To make glass for bottles,** 2% alumina (aluminium oxide) is added to the basic mix.

- **Very cheap glass** is green because it contains small impurities of iron.

▲ Glass is one of the most versatile of all materials – transparent, easily moulded and resistant to the weather. This is why it is used in modern buildings such as this extension to the Louvre in Paris.

- **Metallic oxides** are added to make different colours.

- **Unlike most solids,** glass is not made of crystals and does not have the same rigid structure. It is an amorphous solid.

- **When glass is very hot** it flows slowly like a thick liquid.

> ★ STAR FACT ★
> The person who controls the fires and loads the glass into the furnace is called a teaser.

Radiation

- **Radiation** is an atom's way of getting rid of its excess energy.

- **There are two main kinds** of radiation: electromagnetic and particulate.

- **Electromagnetic radiation** is pure energy. It comes from electrons (see electrons).

- **Particulate radiation** is tiny bits of matter thrown out by the nuclei of atoms.

▲ Exposure to radiation can cause illnesses ranging from nausea to cancer and death. So radioactive materials must be safely guarded.

- **Particulate** radiation comes mainly from radioactive substances (see radioactivity) such as radium, uranium and other heavy elements as they break down.

- **Radiation is measured** in curies and becquerels (radiation released), röntgens (victim's exposure), rads and grays (dose absorbed), rems and sieverts (amount of radiation in the body).

- **Bacteria can stand** a radiation dose 10,000 times greater than the dose that would kill a human being.

- **The Chernobyl nuclear accident** released 50 million curies of radiation. A 20-kilotonne nuclear bomb releases 10,000 times more radiation.

- **The natural radioactivity** of a brazil nut is about six becquerels (0.000000014 curies), which means six atoms in the nut break up every second.

- **The natural background** radiation you receive over a year is about 100 times what you receive from a single chest X-ray.

Water

- **Water is the only substance** that is solid, liquid and gas at normal temperatures. It melts at 0°C and boils at 100°C.

- **Water is at its densest** at 4°C.

- **Ice is much less dense** than water, which is why ice forms on the surface of ponds and why icebergs float.

- **Water is one of the few substances** that expands as it freezes, which is why pipes burst during cold weather.

- **Water has a unique capacity** for making mild solutions with other substances.

- **Water is a compound** made of two hydrogen atoms and one oxygen atom. It has the chemical formula H_2O.

- **A water molecule** is shaped like a flattened V, with the two hydrogen atoms on each tip.

- **A water molecule** is said to be polar because the oxygen end is more negatively charged electrically.

◀ A water molecule has two hydrogen atoms and one oxygen atom in a shallow V-shape.

- **Similar substances** like ammonia (NH_3) are gases to below 0°C.

- **Water stays liquid** until 100°C because pairs of its polar molecules make strong bonds, as the positively charged end of one molecule is drawn to the negatively charged end of another.

▼ Water is found in liquid form, such as rivers, and as a gas in the atmosphere.

Carbon

> ★ STAR FACT ★
> All living things are based on carbon, yet it makes up just 0.032% of the Earth's crust.

- **Pure carbon** occurs in four forms: diamond, graphite, amorphous carbon and fullerenes.

- **Fullerenes** are made mostly artificially, but all four forms of carbon can be made artificially.

- **Diamond** is the world's hardest natural substance.

- **Natural diamonds** were formed deep in the Earth billions of years ago. They were formed by huge pressures as the Earth's crust moved, and then brought nearer the surface by volcanic activity.

- **Graphite** is the soft black carbon used in pencils. It is soft because it is made from sheets of atoms that slide over each other.

- **Amorphous carbon** is the black soot left behind when candles and other objects burn.

▶ The extraordinary hardness of diamonds comes from the incredibly strong tetrahedron (pyramid shape) that carbon atoms form.

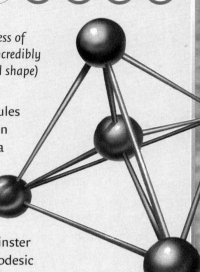

- **Fullerenes** are big molecules made of 60 or more carbon atoms linked together in a tight cylinder or ball. The first was made in 1985.

- **Fullerenes** are named after the architect Buckminster Fuller who designed a geodesic (Earth-shaped) dome.

- **Carbon forms** over one million compounds that are the basis of organic chemistry. It does not react chemically at room temperature. Carbon has the chemical formula C and the atomic number 6. Neither diamond nor graphite melts at normal pressures.

Echoes and acoustics

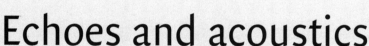

▶ Sydney Opera House in Australia is famous for its stunning design, but some orchestras have complained about its acoustics.

- **An echo** is the reflection of a sound. You hear it a little while after the sound is made.

- **You can only hear** an echo if it comes back more than 0.1 seconds after the original sound.

- **Sound travels** 34 metres in 0.1 seconds, so you only hear echoes from surfaces at least 17 metres away.

- **Smooth hard surfaces** give the best echoes because they break up the sound waves the least.

- **Acoustics** is the study of how sounds are created, transmitted and received.

- **The acoustics** of a space is how sound is heard and how it echoes around that space, whether it is a room or a large concert hall.

- **When concert halls** are designed, the idea is not to eliminate echoes altogether but to use them effectively.

- **A hall with too much echo** sounds harsh and unclear, as echoing sounds interfere with new sounds.

- **A hall without echoes** sounds muffled and lifeless.

- **Even in the best** concert halls, the music can be heard fading after the orchestra stops playing. This delay is called the reverberation time. Concert halls typically have a reverberation time of two seconds. A cathedral may reverberate for up to eight seconds, giving a more mellow, but less clear, sound.

Acids and alkalis

◄ *Citrus fruits such as oranges, lemons and limes have a tart taste because they contain a mild acid, called citric acid. It has a pH of 3.*

- **Mild acids,** such as acetic acid in vinegar, taste sour .

- **Strong acids,** such as sulphuric acid, are highly corrosive. They dissolve metals.

- **Acids** are solutions that are made when certain substances containing hydrogen dissolve in water.

- **Hydrogen atoms** have a single electron. When acid-making substances dissolve in water, the hydrogen atoms lose their electron and become positively charged ions. Ions are atoms that have gained or lost electrons.

- **The strength of an acid** depends on how many hydrogen ions form.

- **The opposite of an acid** is a base. Weak bases such as baking powder taste bitter and feel soapy. Strong bases such as caustic soda are corrosive.

- **A base that dissolves** in water is called an alkali. Alkalis contain negatively charged ions – typically ions of hydrogen and oxygen, called hydroxide ions.

- **When you add an acid** to an alkali, both are neutralized. The acid and alkali react together forming water and a salt.

- **Chemists** use indicators such as litmus paper to test for acidity. Acids turn litmus paper red. Alkalis turn it blue. The strength of an acid may be measured on the pH scale. The strong acid (laboratory hydrochloric) has a pH of 1. The strongest alkali has a pH of 14. Pure water has a pH of about 7 and is neutral – neither acid nor alkali.

> ★ **STAR FACT** ★
> Hydrochloric acid in the stomach (with a pH of between 1 and 2) is essential for digestion.

Solids, liquids and gases

- **Most substances** can exist in three states– solid, liquid or gas. These are the states of matter.

- **Substances** change from one state to another at particular temperatures and pressures.

- **As temperature rises,** solids melt to become liquids. As it rises further, liquids evaporate to become gases.

- **The temperature** at which a solid melts is its melting point.

- **The maximum temperature** a liquid can reach before turning to gas is called its boiling point.

- **Every solid has strength** and a definite shape as its molecules are firmly bonded in a rigid structure.

▲ *Ice melts to water as heat makes its molecules vibrate faster and faster until the bonds between them eventually break.*

- **A liquid has a fixed volume** and flows and takes up the shape of any solid container into which it is poured.

- **A liquid flows** because although bonds hold molecules together, they are loose enough to move over each other, rather like dry sand.

- **A gas** such as air does not have any shape, strength or fixed volume. Its molecules move too quickly for any bonds to hold them together.

- **When a gas cools,** its molecules slow down until bonds form between them to create drops of liquid. This process is called condensation.

Nitrogen

- **Nitrogen** is a colourless, tasteless, odourless, inert (unreactive) gas, yet it is vital to life.

- **Nitrogen is** 78.08% of the air.

- **Nitrogen** turns liquid at −196°C and freezes at −210°C.

- **Liquid nitrogen** is so cold that it can freeze organic substances so quickly they suffer little damage.

- **Food such as cheesecakes** and raspberries are

! NEWS FLASH !
When they die, some people have their bodies frozen with liquid nitrogen in the hope that medical science will one day bring them back to life.

◄ *On average* 100 kg *of nitrate fertilizer are used on every hectare of farmland in the world to replace nitrogen taken from the soil by crops.*

preserved by being sprayed with liquid nitrogen.

- **Compounds of nitrogen** and oxygen are called nitrates.

- **Nitrogen and oxygen** compounds are an essential ingredient of the proteins and nucleic acids from which all living cells are made.

- **Lightning makes** 250,000 tonnes of nitric acid a day. It joins nitrogen and oxygen in the air to make nitrogen oxide.

- **On a long sea dive,** the extra pressure in a diver's lungs makes extra nitrogen dissolve in the blood. If the diver surfaces too quickly the nitrogen forms bubbles, giving 'the bends' which can be painful or even fatal.

Nuclear energy

- **The energy** that binds the nucleus of an atom together is enormous, as Albert Einstein showed.

- **By releasing the energy** from the nuclei of millions of atoms, nuclear power stations and bombs can generate a huge amount of power.

- **Nuclear fusion** is when nuclear energy is released by fusing together small atoms such as deuterium (a kind of hydrogen).

- **Nuclear fusion** is the reaction that keeps stars glowing and gives hydrogen bombs their terrifying power.

- **Nuclear fission** releases energy by splitting the large nuclei of atoms such as uranium and plutonium.

- **To split atomic nuclei** for nuclear fission, neutrons are fired into the nuclear fuel.

- **As neutrons crash** into atoms and split their nuclei, they split off more neutrons. These neutrons bombard other nuclei, splitting off more neutrons that bombard more nuclei. This is called a chain reaction.

- **An atom bomb,** or A-bomb, is one of the two main kinds of nuclear weapon. It works by an explosive, unrestrained fission of uranium-235 or plutonium-239.

- **A hydrogen bomb (H-bomb)** or thermonuclear weapon uses a conventional explosion to fuse the nuclei of deuterium atoms in a gigantic nuclear explosion.

- **The H-bomb** that exploded at Novaya Zemlya in 1961 released 10,000 times more energy than the bombs dropped on Hiroshima, in Japan, in 1945.

Neutron

Nucleus of Uranium atom

▶ *Nuclear fission involves firing a neutron (blue ball) into the nucleus of a uranium or plutonium atom. When the nucleus splits, it fires out more neutrons that split more nuclei, setting off a chain reaction.*

Split nucleus

More neutrons

Spectrum

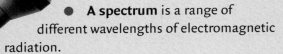

▶ When the beam from a torch passes through a prism it fans out into a rainbow of colours.

- **A spectrum** is a range of different wavelengths of electromagnetic radiation.

- **The white light of sunlight** can be broken up into its spectrum of colours with a triangular block of glass called a prism. The prism is set in a dark room and lit by a shaft of sunlight or similar white light.

- **The prism refracts (bends)** short wavelengths of light more than longer wavelengths, so the light fans out in bands ranging from violet to red.

- **The order of colours** in a spectrum is always the same: red, orange, yellow, green, blue, indigo, violet.

- **Scientists** remember the order of the colours with the first letter of each word in this ancient phrase: 'Richard Of York Gained Battles in Vain'.

- **Infrared** is red light made of waves that are too long for human eyes to see.

- **Ultraviolet** is violet light made of waves that are too short for our eyes to see.

- **Spectroscopy,** or spectral analysis, is the study of the spectrum created when a solid, liquid or gas glows.

- **Every substance** produces its own unique spectrum, so spectroscopy helps to identify substances.

Energy conversion

- **Energy is measured** in joules (J). One joule is the energy involved in moving a force of 1 newton over 1 metre.

- **A kilojoule (kJ)** is 1000 joules.

- **A calorie** was the old measure of energy, but is now used only for food: 1 calorie is 4.187 J; 1 Cal is 1000 calories.

- **For scientists,** 'work' is the transfer of energy. When you move an object, you do work. The work done is the amount of energy (in joules) gained by the object.

- **For scientists, 'power'** is the work rate, or the rate at which energy is changed from one form to another.

- **The power of a machine** is the amount of work it does divided by the length of time it takes to do it.

- **The power of a car's engine** is the force with which the engine turns multiplied by the speed at which it turns.

- **A transducer** is a device for turning an electrical signal into a non-electrical signal (such as sound) or vice versa. A loudspeaker is a transducer.

- **The energy in the Big Bang** was 10^{68} joules. The world's coal reserves are 2×10^{23} joules; a thunderstorm has 1^{14} joules of energy; a large egg has 400,000 joules.

- **When sleeping** you use 60 Cals an hour and 80 Cals when sitting. Running uses 600 Cals. Three hours of reading or watching TV uses 240 Cals. Seven hours' hard work uses about 1000 Cals – or about 10 eggs' worth.

◀ A hydroelectric power station is a device that converts the energy of moving water into electrical energy.

Calcium

- **Calcium** is a soft, silvery white metal. It does not occur naturally in pure form.

- **Calcium** is the fifth most abundant element on Earth.

- **Calcium** is one of six alkaline-earth metals.

- **Most calcium compounds** are white solids called limes. They include chalk, porcelain, enamel on teeth, cement, seashells, the limescale on taps and much more.

- **The word 'lime'** comes from the Latin word for slime.

- **Quicklime** is calcium oxide. It is called 'quick' (Old English for living) because when water drips on it, it twists and swells as if it is alive.

- **Slaked lime** is calcium hydroxide. It may be called 'slaked' because it slakes (quenches) a plant's thirst for lime in acid soils.

- **Calcium has** an atomic number of 20. It has a melting point of 839°C and a boiling point of 1484°C.

- **Limelight** was the bright light used by theatres in the

▲ Calcium is one of the basic building materials of living things. It is one of the crucial ingredients in shell and bone, which is why they are typically white.

days before electricity. It was made by applying a mix of oxygen and hydrogen to pellets of calcium.

- **Calcium adds rigidity** to bones and teeth and helps to control muscles. Your body gets it from milk and cheese.

Oil compounds

◀ This is a propane molecule. The carbon atoms are purple, the hydrogen atoms are grey.

- **Alkanes or paraffins** are a family of hydrocarbons in which the number of hydrogen atoms is two more than twice the number of carbon atoms.

- **Lighter alkanes** are gases like methane, propane and butane (used in camping stoves). All make good fuels.

- **Candles** contain a mixture of alkanes.

- **Alkenes or olefins** are a family of hydrocarbons in which there are twice as many hydrogen atoms as carbon atoms.

- **The simplest alkene** is ethene, also called ethylene (C_2H_4), which is used to make polythene and other plastics such as PVC.

- **Green bananas and tomatoes** are often ripened rapidly in ripening rooms filled with ethene.

- **Ethene** is the basis of many paint strippers.

- **Ethene** can be used to make ethanol – the alcohol in drinks.

▲ In an oil refinery, crude oil is broken down into an enormous range of different hydrocarbons.

- **Hydrocarbons** are compounds made only of carbon and hydrogen atoms. Most oil products are hydrocarbons.

- **The simplest hydrocarbon** is methane, the main gas in natural gas (and flatulence from cows!). Methane molecules are one carbon atom and four hydrogen atoms.

1000 THINGS
YOU SHOULD KNOW ABOUT

BUILDINGS &
TRANSPORT

KEY

 Cars

 Trains

 Planes

 Buildings

 Great monuments

 Boats

Cathedrals

- **Cathedrals** are the churches of Christian bishops.

- **The word cathedral** comes from the Greek word for 'seat', *kathedra*.

- **Cathedrals** are typically built in the shape of a cross. The entrance of the cathedral is at the west end.

- **The long end** of the cross is the nave.

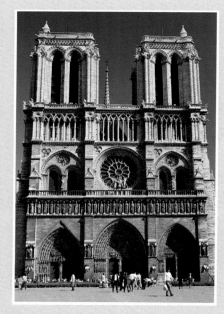

▲ *Paris's famous Notre Dame dates from 1163, but was remodelled in the 1840s by Viollet-le-Duc.*

- The short end, containing the altar and choir stalls, is called the apse. The arms of the cross are called transepts.

- **Canterbury and a few other cathedrals** were built as early as the 6th-century AD. Most European cathedrals were built between 1000 and 1500.

- **Medieval cathedrals** were often built in the 'gothic' style with soaring, pointed arches, tall spires and huge pointed arch windows.

- **Many gothic cathedrals** have beautiful stained-glass windows. Chartres Cathedral in France has 176.

- **The modern Crystal Cathedral** in California is star-shaped and made of glass for TV broadcasts.

- **The world's tallest** cathedral spire is in Ulm, Germany, which is 160.9 m tall.

- **The world's smallest** cathedral is Christ Church in Highlandville, Missouri, USA which is just 4.2 by 5.2 m and holds only 18 people.

The first railways

- **Railways** were invented long before steam power.

- **The Diolkos** was a 6 km long railway that transported boats across the Corinth isthmus in Greece in the 6th-century BC. Trucks pushed by slaves ran in grooves in a limestone track.

- **The Diolkos** ran for over 1300 years until AD 900.

- **Railways** were revived in Europe in the 14th century with wooden trackrails to guide horse and hand carts taking ore out of mines.

- **In the 1700s,** English ironmakers began to make rails of iron. First the rail was wood covered in iron. Then the whole rail was made of iron. Iron wheels with 'flanges' (lips) ran inside the track.

▶ *The Stephensons' 'Rocket' was the most famous early locomotive, winning the first locomotive speed trials at Rainhill in England in 1829.*

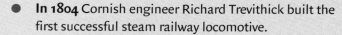

- **In 1804** Cornish engineer Richard Trevithick built the first successful steam railway locomotive.

- **Trevithick's** engine pulled a train of five wagons with 9 tonnes of iron and 70 men along 15 km of track at the Pendarren ironworks in Wales.

- **On 27 September 1825** George and Robert Stephenson opened the world's first steam passenger railway, the Stockton and Darlington in England.

- **The gauge (trackwidth)** used for the Stockton and Darlington was 1.44 m, the same length as axles on horse-wagons. This became the standard gauge in the USA and much of Europe.

- **The English-built 'Stourbridge Lion'** was the first full-size steam locomotive to run in the USA. It ran on wooden track in Pennsylvania in 1829.

Controlling a plane

- **A plane** is controlled in the air by moving hinged flaps on its wings and tail.
- **Changing pitch** is when the plane goes nose-up or nose-down to dive or climb.
- **Rolling** is when the plane rolls to one side, dipping one wing or the other.
- **Yawing** is when the plane steers to the left or right like a car.
- **Pitch** is controlled by raising or lowering hinged flaps on the rear wings called elevators.
- **To pitch up** in a small or simple plane, the pilot pulls back on the control column to raise the elevators.
- **Rolling** is controlled by large hinged flaps on the wings called ailerons.

◄ In old-fashioned planes, the pilot controlled the flaps manually by moving a control stick linked to the flaps by cables. In modern planes, the flaps are controlled automatically via electric wires (fly-by-wire) or laser beams (fly-by-light). The flight deck of this plane from 30 years ago has lots of dials to help the pilot. Modern planes have 'glass cockpits', which means they have computer screens.

- **To roll left,** the pilot pushes the control column to the left, which raises the aileron on the left wing and lowers it on the right.
- **Yawing** is controlled by the vertical hinged flap on the tail called the rudder.
- **To yaw left,** the pilot pushes the foot-operated rudder bar forward with his left foot, to swing the rudder left.

Trucks

- **Trucks or lorries** can weigh 40 tonnes or more – maybe as much as 50 cars.
- **Cars** are powered by petrol engines; trucks are typically powered by diesel engines.
- **Cars** typically have five forward gears; trucks often have as many as 16.
- **Cars** typically have four wheels. Trucks often have 12 or 16 to help spread the load.
- **Many trucks** have the same basic cab and chassis (the framework supporting the body). Different bodies are fitted on to the chassis to suit the load.
- **Some trucks** are in two parts, hinged at the join. These are called articulated trucks.
- **The cab and engine** of an articulated truck is called the tractor unit. The load is carried in the trailer.
- **In Australia** one tractor may pull several trailers in a 'road-train' along the long, straight desert roads.

▲ This 14-wheel tanker is articulated, and the tank and tractor unit can be separated.

- **The longest truck** is the Arctic Snow Train, first made for the US army. This is 174 m long and has 54 wheels.
- **Large trucks** are sometimes called 'juggernauts'. Juggernaut was a form of the Hindu god Vishnu who rode a huge chariot, supposed by Europeans to crush people beneath its wheels.

Mosques

- **The word 'mosque'** comes from the Arabic *masjid*, which means a place of kneeling.

- **Mosques** are places where Muslims worship. Muslims can worship anywhere clean. So a mosque can be just a stick in the sand to mark out a space for praying or even just a mat. But many mosques are beautiful buildings.

- **Cathedral or Friday** mosques are large

▲ *Typically minarets on mosques have onion-shaped roofs.*

mosques big enough to hold all the adult Muslims in the area.

- **Typically** mosques have a courtyard surrounded by four walls called iwans. There is often a fountain or fuawara at the centre for ceremonial washing.

- **The mihrab** is a decorative niche on the inner wall closest to the city of Mecca that muslims face when praying.

- **The mimbar** is the stone or wooden pulpit where the Imam leads the people in prayer.

- **Most mosques** have two to six tall pointed towers which are called minarets.

- **On each minaret** is a balcony from which muezzin (criers) call the faithful to prayer five times a day.

- **Minarets** may have been inspired by the Pharos, the famous lighthouse at Alexandria in Egypt.

- **There are** no paintings or statues in mosques, only abstract patterns, often made of tiles.

Motorbikes

▲ *Many enthusiasts still ride motorbikes like this from the 1930s when motorcycling was in its heyday.*

- **The first petrol engine** motorbike was built by Gottlieb Daimler in Germany in 1885.

- **Most small motorbikes** have a two-stroke petrol engine, typically air-cooled and quite noisy.

- **Most larger motorbikes** have four-stroke petrol engines, typically water-cooled.

- **On most bikes** the engine drives the rear wheel via a chain; on some the drive is via a shaft.

- **A trail bike** is a bike with a high mounted frame and chunky tyres, designed to be ridden over rough tracks.

- **Motorbikes** with engines smaller than 50cc may be called mopeds, because in the past they had pedals as well as an engine.

- **The biggest motorbikes** have engines of around 1200cc.

- **The most famous** motorbike race in the world is the Isle of Man TT, or Tourist Trophy.

- **In Speedway,** bikes with no brakes or gears are raced round dirt tracks.

> ★ STAR FACT ★
> The first motorbike was steam powered, and built by the Michaux brothers in 1868.

Building of the railways

- **On the 15th September 1830** the world's first major passenger railway opened between Liverpool and Manchester.

- **At the opening** of the Liverpool and Manchester railway, government minister William Huskisson was the first railway casualty, killed by the locomotive.

- **In 1831** the Best Friend started a regular train service between Charleston and Hamburg, South Carolina in the USA.

- **By 1835** there were over 1600 km of railway in the USA.

- **In Britain** railway building became a mania in the 1840s.

- **Hundreds of acts** of parliament gave railway companies the powers to carve their way through cities

◀ *The new railways often meant demolishing huge strips through cities to lay track into stations.* and the countryside .

- **By the late 1840s** Great Western Railway trains were able to average well over 100 km/h all the way from London to Exeter via Bristol, completing the 300 odd km journey in under 4 hours.

- **On 10th May 1869** railways from either side of the United States met at Promontory, Utah, giving North America the first transcontinental railway.

- **The British** built 40,000 km of railways in India in the 1880s and 90s.

- **Vast armies of men** worked on building the railways – 45,000 on the London and Southampton railway alone.

Early boats

- **The Aborigines** arrived by boat in Australia at least 50,000 years ago, as cave paintings show.

- **The oldest remains** of boats are 8000 or so years old, like the 4 m canoe found at Pesse in Holland.

- **Many early canoes** were simply hollowed out logs, called dugouts.

- **The islands of the Pacific** were probably colonized by dugouts, stabilized by attaching an outrigger (an extra float on the side) or adding extra hulls.

- **Lighter canoes** could be made by stretching animal skins over a light frame of bent wood.

- **4000 years ago,** Ancient Egyptians were making ships over 30 m long by interlocking planks of wood and lashing them together with tough grass rope.

- **By 1000 BC,** Phoenician traders were sailing into the Atlantic in ships made of planks of Cedar trees.

▲◀ *Outriggers (above) were used to colonize the Pacific tens of thousands of years ago. But many early boats like the Welsh coracle (left) were made by stretching animal skins over a wooden frame.*

- **All early boats** were driven by hand – using either a pole pushed along the bottom, or a paddle.

- **Sails** were first used 7000 years ago in Mesopotamia on reed trading boats.

- **The first known picture** of a sail is 5500 years old and comes from Egypt.

Castles

- **Castles** were the fortress homes of powerful men such as kings and dukes in the Middle Ages. The castle acted as a stronghold for commanding the country around it. They were also barracks for troops, prisons, armouries and centres of local government.

- **The first castles,** in the 11th century, had a high earth mound called a motte, topped by a wooden tower. Around this was an enclosure called a bailey, protected by a fence of wooden stakes and a ditch.

- **From the 12th century** the tower was built of stone and called a keep or donjon. This was the last refuge in an attack. Soon the wooden fences were replaced by thick stone walls and strong towers.

- **Walls and towers** were topped by battlements – low walls with gaps for defenders to fire weapons from.

- **Castles evolved** from a simple square tower to elaborate rings of fortifications. The entrance or gatehouse was equipped with booby traps.

- **From the 13th century** the ditch around the castle was often dug deep, filled with water and called a moat. This stopped enemies from sapping (digging under the walls).

- **Many castles** in England and Wales were first built by the Norman conquerors in the 11th century, including Windsor Castle which dates from 1070.

- **In early castles** there was just a single great hall. Later castles had extra small rooms for the lord and lady, such as the solar, which had windows.

- **The gong farmer** was the man who had to clean out the pit at the bottom of the garderobe (toilet shute).

- **An attack on a castle** was called a siege and could last many months, or even years.

▼ This shows some of the main features of a medieval castle. Very few castles had all of these features. Innovations were constantly being added over the centuries as attackers came up with better ways of breaching the defences and defenders found ways to hold them at bay.

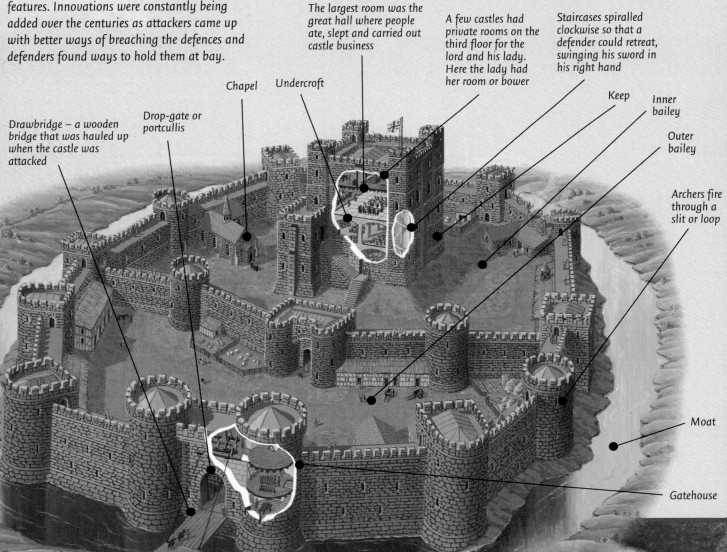

The largest room was the great hall where people ate, slept and carried out castle business

A few castles had private rooms on the third floor for the lord and his lady. Here the lady had her room or bower

Staircases spiralled clockwise so that a defender could retreat, swinging his sword in his right hand

Chapel

Undercroft

Keep

Inner bailey

Drawbridge – a wooden bridge that was hauled up when the castle was attacked

Drop-gate or portcullis

Outer bailey

Archers fire through a slit or loop

Moat

Gatehouse

Galleons

◄ *In this replica of a ship from around 1600, the tall stern castle where the captain lived is clearly visible.*

- **Galleons** were huge sailing ships that were built from the mid 1500s onwards.

- **Galleons** had tall structures called castles at either end.

- **The front castle** was called the forecastle or fo'c'sle (said folk-sel).

- **The rear castle** was called the stern castle and it housed elaborate living quarters for the officers.

The crew lived in crowded rows below deck.

- **The stern of the ship** was often ornately carved and painted in gold and bright colours.

- **A galleon** had three tall masts – the foremast, the mainmast and the mizzenmast at the rear.

- **Galleons** were both warships and trading boats and carried vast numbers of cannon and troops.

- **In the 1500s,** enormous Spanish galleons crossed the Atlantic carrying gold from the Americas to Spain.

- **When the Spanish Armada** of galleons was sent to attack England in 1588, they were outfought by the smaller but faster and lighter English ships, despite their heavy armament.

> ★ STAR FACT ★
> The biggest Spanish galleons weighed well over 1000 tonnes.

The Colosseum

- **The Colosseum** was a huge stone sports arena built in Ancient Rome.

- **It was 189 m long,** 156 m wide, and 52 m high.

- **It held** up to 73,000 spectators who entered through 80 entrances. Each spectator had a ticket corresponding to one of the 76 arcades.

- **The sports** included fights between gladiators with swords, nets and other weapons who fought to the death, or fought against lions and other wild beasts.

- **Counterweighted** doors allowed 64 wild beasts to be released from their cages simultaneously.

- **It took just eight years** – from around AD70 to AD78 – to build the Colosseum.

- **To build it** the Romans brought almost a quarter of a million tonnes of stone by barge from quarries 20 km outside Rome.

- **During construction,** a cart carrying a tonne of stone would have left the riverside wharves every seven

▲ *The Colosseum in Rome was one of the greatest buildings of the ancient world.*

minutes on the 1.5 km journey to the site.

- **A huge awning** called a velarium, supported by 240 masts, protected the arena from bad weather.

- **Its opening was celebrated** by spectacular games lasting 100 days.

Tractors

- **Tractors** are motor vehicles used mostly on farms for pulling or pushing machines such as ploughs.

- **Tractors** are also used in factories, by the army, for logging and for clearing snow.

- **Steam-powered tractors** called 'traction engines' were introduced in 1834 by Walter Hancock.

- **Traction-engines** could pull up to 40 ploughs at once via long chains, but they were too cumbersome to be practical in hilly country.

- **Petrol-powered tractors** were introduced in the 1890s but they were not powerful enough for most farm work.

- **The first all-purpose** farm tractors appeared in the 1920s and soon began to replace horses and oxen.

▲ *Modern tractors can be adapted to operate all kinds of devices, including this haybailer, via the power take-off.*

- **Ploughs and other equipment** are drawn along via the drawbar on the back of the tractor.

- **The power take-off** or PTO allows the tractor's engine to power equipment such as potato diggers.

- **25% of all tractors** are in the USA.

> ★ STAR FACT ★
> There are now well over 16 million tractors around the world.

The Age of Sail

- **East Indiamen** were ships that carried ivory, silks and spices to Europe from India, China and the East Indies.

- **By 1800** Indiamen could carry 1000 tonnes of cargo.

- **Packet ships** were ships that provided a regular service across the Atlantic between Europe and the USA – no matter whether they had a full load or not.

- **The first packet service** began in 1818 with the Black Ball Line from Liverpool to New York. Red Star and Swallowtail followed.

◀ *The tea clippers of the mid-1800s were the pinnacle of sailing ship technology, able to carry thousands of tonnes of tea at speeds of 20 knots or more.*

- **Clippers** were the big, fast sailing ships that 'clipped off' time as they raced to get cargoes of tea from China and India back first to markets in Europe and the USA.

- **Clippers** had slender hulls, tall masts and up to 35 sails.

- **Clippers could reach** speeds of 30 knots (55 km/h). Many could sail from New York, round South America, to San Francisco in under 100 days.

- **In 1866** the clippers *Taeping*, *Serica* and *Ariel* raced 25,700 km from Foochow in China to London in 99 days.

- **Canadian Donald Mckay** of Boston, Massachusetts, was the greatest clipper builder. His ship the *Great Republic* of 1853 was the biggest sailing ship of its time, over 100 m long and able to carry 4000 tonnes of cargo.

- **A famous clipper,** the *Cutty Sark*, built in 1869 is preserved at Greenwich, London. Its name is Scottish for 'short shirt' and comes from the witch in Robert Burns's poem *Tam O'Shanter*.

Parachutes, hang-gliders

- **Leonardo da Vinci** drew a design for a parachute around 1500.

- **The first human parachute drop** was made by Jacques Garnerin from a balloon over Paris in 1797.

- **Folding parachutes** were used first in the USA in 1880.

- **Early parachutes** were made of canvas, then silk. Modern parachutes are made of nylon, stitched in panels to limit tears to a small area.

- **Until recently** parachutes were shaped like umbrellas; now most are 'parafoils' – wing-shaped.

- **Drogues** are parachutes thrown out by jet planes and high-speed cars to slow them down.

- **German Otto Lilienthal** flew his own canvas and wood hang-gliders in the 1890s.

- **Modern hang-gliding** began with the fabric delta

◀ *Parachutes really came into their own with the massive parachute drops of troops into battle during World War 2.*

(triangular) wing design developed by the American Francis Rogallo in the 1940s.

- **Today's hang-gliders** are made by stretching nylon over a very light and strong nylon and kevlar frame. They combine long aeroplane-like wings and a double skin of fabric to achieve very long flights.

- **The first hang-gliders** achieved a glide ratio of 1:2.5 – that is, they travelled only 2.5 m forward for every 1 m that they dropped. Today's hang-gliders give glide ratios of 1:14 or better.

Skyscrapers

- **The first skyscrapers** were built in Chicago and New York in the 1880s.

- **A crucial step** in the development of skyscrapers was the invention of the fast safety lift by US engineer Elisha Otis (1811-61) in 1857.

- **The Home Insurance Building** in Chicago, built in 1885, was one of the first skyscrapers.

- **In buildings** over 40-storeys high, the weight of the building is less important in terms of strength than the wind load (the force of the wind against the building).

- **The Empire State Building,** built in New York in 1931, was for decades the world's tallest building at 381 m.

- **The tallest building** in America is the 442 m-high Sears Tower in Chicago, built in 1974.

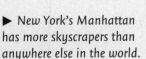

! NEWS FLASH !
The Landmark Tower in Kowloon, Hong Kong, could be 574 m high.

▶ *New York's Manhattan has more skyscrapers than anywhere else in the world.*

- **The world's tallest building** is the 452 m Petronas twin towers in Kuala Lumpur, Malaysia.

- **If the Grollo Tower** is built in Melbourne Australia it will be 560 m high.

- **An early concept** design has been presented for a Tokyo building measuring 4000 m!

St Sophia

- **Saint or Hagia Sophia** was built as an ancient Christian cathedral in Istanbul in Turkey (Roman 'Byzantium').

- **Hagia Sophia** is Greek for 'holy wisdom'.

- **St Sophia** was built between 532 and 537 AD, when Istanbul was called Constantinople.

- **The great Byzantine** emperor Justinian I ordered the building of St Sophia after a fire destroyed an earlier church on the site.

- **St Sophia** was designed by the brilliant Byzantine architects Anthemius of Tralles and Isidorus of Miletus.

- **The main structure** is a framework of arches and vaults (high arched ceilings).

- **The dome** is 31 m across and 56 m high. It is supported by four triangular brick pillars called pendentives.

- **St Sophia is decorated** with marble veneers and mosaics of Mary, Christ, angels and bishops.

- **In 1453** the Ottoman Turks took the city and converted St Sophia to a mosque. They plastered over the mosaics and added four minarets on the outside.

- **In 1935** St Sophia was converted into a museum and the mosaics were uncovered.

▼ *Istanbul's St Sophia is one of the world's oldest cathedrals and one of the most remarkable pieces of ancient architecture.*

Viking longships

- **Viking longships** were the long narrow boats built by the Viking warriors of Scandinavia around 1000 AD.

- **A virtually** intact ship was found in 1880 at Gokstad near Oseberg in Norway. A king was buried inside it.

- **Longships** were very fast, light and seaworthy.

▲ *The Gokstad ship dates from c. AD 900.*

They carried the Viking explorer Leif Ericsson all the way across the Atlantic to North America.

- **In 1893** a replica of the Gokstad ship sailed across the Atlantic in 28 days.

- **Some Viking longships** were up to 30 m long and 17 m across and could carry 200 warriors.

- **At sea** longships relied on a single square sail made up from strips of woollen cloth; on rivers and in calm they used oar-power, with 20-30 rowers on each side.

- **Their ships' shallow build** allowed the Vikings to make raids far up shallow rivers.

- **The prow curved** up to a carved dragon head.

- **Longships** were for carrying warriors on raids, but the Vikings also built wider, deeper 'knorrs' for trade and small rowing boats called 'faerings'.

- **The Viking** ships were very stable because they had a keel, a wooden board about 17 m long and 45 cm deep, along the bottom of the boat.

Diesel trains

> ★ STAR FACT ★
> In 1987, a British Rail diesel train set a world diesel speed record of nearly 240 km/h.

- **The diesel engine** was invented by Rudolf Diesel in 1892 and experiments with diesel locomotives started soon after. The first great success was the *Flying Hamburger* which ran from Berlin to Hamburg in the 1930s at speeds of 125 km/h. Diesel took over from steam in the 1950s and 1960s.

- **Diesel locomotives** are really electric locomotives that carry their own power plant. The wheels are driven by an electric motor which is supplied with electricity by the locomotive's diesel engine.

- **The power output of a diesel** engine is limited, so high-speed trains are electric – but diesels can supply their own electricity so need no trackside cables.

- **There are two other kinds** of diesel apart from diesel-electrics: diesel-hydraulic and diesel-mechanical.

- **In diesel-hydraulics,** the power from the diesel engine is connected to the wheels via a torque converter, which is a turbine driven round by fluid.

- **In diesel-mechanicals,** the power is transmitted from the diesel engine to the wheels via gears and shafts. This only works for small locomotives.

- **Diesel locomotives** may be made up from one or more separate units. An A unit holds the driver's cab and leads the train. A B unit simply holds an engine.

- **A typical diesel** for fast, heavy trains or for trains going over mountains may consist of one A unit and six B units.

- **The usual maximum** power output from a single diesel locomotive is about 3500–4000 horsepower. In Russia, several 3000-horsepower units are joined together to make 12,000-horsepower locomotives.

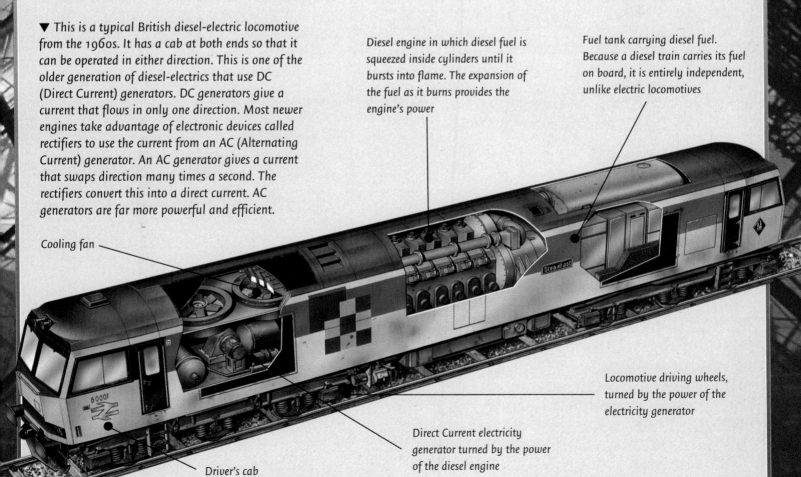

▼ This is a typical British diesel-electric locomotive from the 1960s. It has a cab at both ends so that it can be operated in either direction. This is one of the older generation of diesel-electrics that use DC (Direct Current) generators. DC generators give a current that flows in only one direction. Most newer engines take advantage of electronic devices called rectifiers to use the current from an AC (Alternating Current) generator. An AC generator gives a current that swaps direction many times a second. The rectifiers convert this into a direct current. AC generators are far more powerful and efficient.

Diesel engine in which diesel fuel is squeezed inside cylinders until it bursts into flame. The expansion of the fuel as it burns provides the engine's power

Fuel tank carrying diesel fuel. Because a diesel train carries its fuel on board, it is entirely independent, unlike electric locomotives

Cooling fan

Locomotive driving wheels, turned by the power of the electricity generator

Driver's cab

Direct Current electricity generator turned by the power of the diesel engine

Supersonic planes

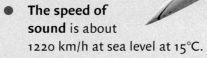

▶ The Soviet Tupolev Tu-144 of 1968 was the very first supersonic jetliner, but the Anglo-French Concorde of 1969 is the only one that is still flying.

- **Supersonic planes** travel faster than the speed of sound.
- **The speed of sound** is about 1220 km/h at sea level at 15°C.
- **Sound travels** slower higher up, so the speed of sound is about 1060 km/h at 12,000 m.
- **Supersonic** plane speeds are given in Mach numbers. These are the speed of the plane divided by the speed of sound at the plane's altitude.
- **A plane flying** at 1500 km/h at 12,000 m, where the speed of sound is 1060 km/h, is at Mach 1.46.
- **A plane flying** at supersonic speeds builds up shock waves in front and behind because it keeps catching up and compressing the sound waves in front of it.
- **The double shock waves** create a sharp crack called a sonic boom that is heard on the ground. Two booms can often be heard a second or two apart.
- **In 1947** Chuck Yeager of the USAF made the first supersonic flight in the Bell X-1 rocket plane. The X-15 rocket plane later reached speeds faster than Mach 6. Speeds faster than Mach 5 are called hypersonic.
- **The first jet plane** to fly supersonic was the F-100 Super Sabre fighter of 1953. The first supersonic bomber was the USAF's Convair B-58 Hustler of 1956.

Famous castles

▲ Eilean Donan in Scotland was used in the film 'Braveheart'.

- **The Arab Gormaz Castle** in the Castile region of Spain is Western Europe's biggest and oldest castle. It was started in 956 AD by Caliph Al-Hakam II. It is 1 km round and over 400 m across.
- **Bran Castle** in Romania was built in 1212 by Teutonic (German) knights. In the 1450s, it was one of the castles of Vlad the Impaler, the original Count Dracula.
- **Windsor Castle** in England has been one of the main homes of English kings and queens since the 1100s.
- **Malbork Castle** near Gdansk in Poland is the biggest medieval castle, built by the Teutonic knights in 1309.
- **Germany's Neuschwanstein** castle was started in 1869 by 'Mad' King Ludwig II. Its name means 'new swan castle' after Swan Knight in Wagner's opera Lohengrin.
- **Neuschwanstein** is the model for the castle in Walt Disney's cartoon film Sleeping Beauty.
- **15th century Blarney Castle** in Ireland is home to the famous 'Blarney stone'. Kissing the stone is said to give people the 'gift of the gab' (fast and fluent talk).
- **Tintagel Castle** in Cornwall is where legend says the mythical King Arthur was conceived.
- **Colditz Castle** in Germany was a notorious prison camp in World War 2.
- **Carcassone** in southern France is a whole town fortified by Catholics c.1220 against those of the Cathar religion.

Golden Gate Bridge

- **The Golden Gate Bridge** spans the entrance to San Francisco Bay in California, USA.

- **The Golden Gate Bridge** is one of the world's largest suspension bridges.

- **The total length** of the bridge is 2737 m.

◀ *San Francisco's Golden Gate Bridge is one of the most beautiful – and busiest – bridges in the world.*

- **The towers** at either end are 340 m high. The suspended span between the towers is 1280 m, one of the longest in the world.

- **The roadway is** 67 m above the water, although this varies according to the tide

- **The Golden Gate Bridge** was designed by Joseph Baerman Strauss and built for $35 million, a third of the original cost estimate.

- **The bridge** was opened to traffic on 27 May 1937.

- **The bridge** carries two six-lane highways and pedestrian paths.

- **Unusually,** the bridge is double-deck, carrying traffic one way on the upper deck and the other way on the lower deck.

Junks

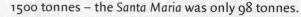

- **Junks** are wooden sailing boats that have been used in China and the Far East for thousands of years.

- **Typical junks** have a broad flat bow (front).

- **Typical junks** have a broad, high stern castle (rear).

- **Junks have lugsails**. These are triangular sails arranged almost in line with the boat, allowing the boat to sail almost into the wind – unlike the square sails of early European boats which only worked with the wind behind.

- **The Chinese** are believed to have invented lugsails.

- **By the 1400s,** the Chinese were building junks 150 m long and almost 100 m wide – much bigger than any sailing ship yet built in Europe.

- **In 1405** the Chinese admiral Cheng Ho led a fleet of exploration through the Indian Ocean. His fleet consisted of 62 large junks and 255 small junks. Each of the large junks was gigantic compared with Columbus's *Santa Maria*. The biggest junk was over 1500 tonnes – the *Santa Maria* was only 98 tonnes.

- **Between 1790 and 1810,** traders in the South Seas were terrorized by vast fleets of pirate junks, led by the female pirate Cheng I Sao.

- **Nowadays** most junks are motorized.

- **Junks** are often moored permanently as homes.

▶ *Most sailing junks have two or three masts. Some have more. The sails are made from cotton and supported by bamboo struts.*

Fighters of World War II

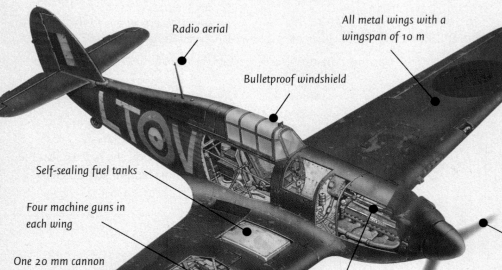

Radio aerial

All metal wings with a wingspan of 10 m

Bulletproof windshield

Self-sealing fuel tanks

Four machine guns in each wing

One 20 mm cannon in each wing

1030 hp Rolls-Royce Merlin engine capable of powering the plane to over 520 km/h

Three-blade propeller

◀ Along with the Spitfire, the Hurricane was the mainstay of the British defence against the German air invasion in the Battle of Britain in 1940. Hurricanes and Spitfires would have spectacular aerial 'dogfights' with the Me 109s and FW 190s escorting the German bombers. The sturdy Hurricane proved a highly effective fighter plane.

★ STAR FACT ★
The 'Night Witches' were a crack Soviet squadron of female fighter pilots who flew night raids on the Germans in the Caucasus Mountains.

▲ In the dogfights of the Battle of Britain, the Spitfire's 650 km/h top speed and amazing agility proved decisive.

- **World War II** fighter planes were sleek monoplanes (single-winged aircraft) very different from the biplanes of World War I. Many were developed from racing machines of the 1920s and 1930s.

- **The most famous** British plane was the Supermarine Spitfire. This was developed from the S.6B seaplane which won the coveted Schneider trophy in the late 1920s.

- **The most famous American** fighter was the North American P-51 Mustang, which could reach speeds of over 700 km/h and had a range of 1700 km. It was widely used as an escort for bombers.

- **In the Pacific** US planes like the Grumman Wildcat and Hellcat fought against the fast and highly manoeuvrable Japanese Mitsubishi A6M 'Zero'.

- **The most famous German** fighter was the fast, agile Messerschmitt Bf 109. This was the plane flown by German ace Erich Hartmann, who shot down 352 enemy planes. Over 33,000 Me 109s were made during the war.

- **Me 109s** were nicknamed by model. The Bf109E was the Emil. The Bf109G was the Gustav.

- **The German Focke-Wulf 190** had a big BMW radial engine that enabled it to climb over 1100 km a minute.

- **The most famous Soviet** fighter was the MiG LaGG-3 Interceptor, flown by Soviet air ace Ivan Kozhedub who shot down 62 German planes.

- **The British Hawker Hurricane** was more old-fashioned, slower and less famous than the Spitfire, but its reliability made it highly effective. Hurricanes actually destroyed more enemy aircraft than Spitfires.

The Benz

- **Karl Benz (1844-1929)** and his wife Berta were the German creators of the first successful petrol-driven car in 1885.

- **At the age of 16,** before the first petrol-powered cart was built, Karl Benz dreamed of 'a unit that could supplant the steam engine'.

- **Karl and Berta Benz** began developing their car in the 1870s, while trying to earn money from a tin-making business. By 1880, they were so penniless they could barely afford to eat everyday.

- **Their first car** was a small tricycle, but had a water-cooled four-stroke engine with electric ignition.

▶ *The Benz tricycle was the world's first successful petrol-driven car.*

- **Karl Benz** tested the car round the streets of their home town of Mannheim after dark to avoid scaring people.

- **Berta Benz** recharged the car's battery after each trip by pedalling a dynamo attached to her sewing machine.

- **Berta Benz** secretly made the first long, 100 km journey in the car, without Karl knowing. On the journey to her home in Pforzheim, she had to clear the fuel-line with a hairpin and use her garters to cure an electrical short.

- **The Benz** caused a stir when shown at the Paris World Fair in 1889.

- **Backed by F. von Fischer** and Julian Ganss, Benz began making his cars for sale in 1889 – the first cars ever made for sale.

- **By 1900** Benz was making over 600 cars a year.

Stonehenge

- **Stonehenge** is an ancient stone monument in southern England made from circles of huge rough-cut stones.

- **The main circle of stones** or 'sarsens' is a ring of 30 huge upright stones, joined at the top by 30 lintels. Many of these sarsens have now fallen or been looted.

- **Inside the sarsen ring** are five 'doorways' called trilithons, each made with three gigantic stones, weighing up to 40 tonnes each.

- **In between** are rings of smaller bluestones.

- **The bluestones** come from the Preseli Hills in Wales, 240 km away. Archeologists puzzle over how they were carried here.

▶ *Experiments have shown a team of 150 people could haul the stones upright, but dragging them to the site on greased wooden sleds must have been a huge undertaking.*

- **At the centre** is a single tall stone, called the Heel Stone.

- **Stonehenge was built** in three phases between 2950 BC and 1600 BC, starting with just a huge earth ring.

- **Archaeologists** believe it was a gathering place and a site for religious ceremonies for Bronze Age people.

- **Newman, Thom and Hawkins** have shown the layout of the stones ties in with astronomical events.

- **At sunrise on midsummer's day** (the solstice), the sun shines directly along the avenue to the Heel Stone.

Warplanes

◀ Pilots flying modern jets fly at supersonic speeds aided by laser-guided weapons, night-vision goggles and other high-tech equipment.

- **The 870 km/h** German Messerschmitt Me 262 was the first jet fighter. It had straight wings like propeller planes.

- **The Lockheed Shooting Star** was the first successful US jet fighter.

- **The Korean War** of the 1950s saw the first major combat between jet fighters. Most now had swept-back wings, like the Russian MiG-15 and the US F-86 Sabre.

- **In 1954** Boeing introduced the B-52 Superfortress, still the USAF's main bomber because of its huge bomb-carrying capacity.

- **In the 1950s** aircraft began flying close to the ground to avoid detection by radar. On modern planes like the Lockheed F-111 a computer radar system flies the plane automatically at a steady height over hills and valleys.

- **The Hawker Harrier** of 1968 was the only successful 'jump jet' with swivelling jets for vertical take-off (VTOL).

- **Airborne Early Warning** systems (AEWs) detect low-flying aircraft. To evade them, the Americans began developing 'stealth' systems like RCS and RAM.

- **RCS** or Radar Cross Section means altering the plane's shape to make it less obvious to radar. RAM (Radar Absorbent Material) is a coating that doesn't reflect radar.

- **In 1988** the US unveiled its first 'stealth' bomber, the B-2, codenamed Have Blue. The F117 stealth fighter followed.

- **The 2500 km/h Russian Sukhoi S-37** Berkut ('golden eagle') of 1997 uses Forward Swept Wings (FSW) for maximum agility, rather than stealth technology.

Gears

▲ Gears are used in a huge range of machines, from watches to motorbikes, for transmitting movement from one shaft to another.

- **Gears** are pairs of toothed wheels that interlock and so transmit power from one shaft to another.

- **The first gears** were wooden, with wooden teeth. By the 6th century AD, wooden gears of all kinds were used in windmills, watermills and winches.

- **Metal gears** appeared in 87 BC and were later used for clocks. Today, all gears are metal and made on a 'gear-hobbing' machine.

- **Simple gears** are wheels with straight cut teeth.

- **Helical gears** have slanting rather than straight teeth, and run smoother. The gears in cars are helical.

- **Bevel or mitre gears** are cone-shaped, allowing shafts to interlock at right angles.

- **In worm gears** one big helical gear interlocks with a long spiral gear on a shaft at right angles.

- **Planetary gears** have a number of gear wheels called spurs or planets that rotate around a central 'sun' gear.

- **In a planetary gear** the planet and sun gear all rotate within a ring called an annulus.

> ★ STAR FACT ★
> Automatic gearboxes in cars use planetary or epicyclic gears.

Arches

- **Before arches,** door openings were just two uprights spanned by a straight piece of wood or stone called a lintel.

- **Arches** replace a straight lintel with a curve. Curved arches can take more weight than a lintel, because downwards weight makes a lintel snap in the middle.

- **The Romans** were the first to use round arches, so round arches are called Roman or Romanesque arches.

- **Roman arches** were built from blocks of stone called voussoirs. They were built up from each side on a semi-circular wooden frame (later removed). A central wedge or keystone is slotted in at the top to hold them together.

- **The posts** supporting arches are called piers. The top of each post is called a springer.

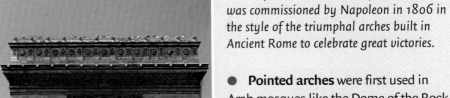

◄ *The famous Arc de Triomphe in Paris was commissioned by Napoleon in 1806 in the style of the triumphal arches built in Ancient Rome to celebrate great victories.*

- **Pointed arches** were first used in Arab mosques like the Dome of the Rock in the 7th century.

- **Pointed arches** were brought to Europe from the Middle East by Crusader knights in the 1100s and put in churches to become part of the gothic style.

- **Horseshoe arches** are used in islamic buildings all round the world.

- **The sides** of an ogee arch are S-shaped. Tudor arches are flattened ogees.

- **The world's biggest free-standing arch** is the Gateway to the West arch in St Louis, Missouri, USA. Completed in 1965, this arch is 192 m high and 192 m across.

Record breaking trains

- **The fastest steam train** ever was by Gresley-designed steam engine *Mallard* pulling seven coaches on 3 July 1938, (see steam locomotives). It travelled at 201 km/h.

- **The most powerful** steam loco was the US Virginian Railway's No.700. It pulled with a force of over 90,000 kg.

- **The heaviest trains** ever pulled by a single locomotive were 250-truck trains that ran on the Erie Railroad in the USA from 1914 to 1929. They weighed over 15,000 tonnes.

- **The longest train** was a 7.3 km 660-truck train that ran from Saldanha to Sishen in South Africa on 26 Aug 1989.

- **The longest passenger train** was a 1732 m 70-coach train from Ghent to Ostend in Belgium on 27 April 1991.

- **The fastest diesel train** was 248 km/h by a British Rail Intercity 125 between Darlington and York on 1 November 1987.

- **The fastest scheduled service** is the Hiroshima-Kokura

bullet train in Japan which covers 192 km in 44 minutes at an average 261.8 km/h.

- **The TGV** from Lille to Roissy in France covers 203 km in 48 mins at an average 254.3 km/h.

- **The fastest train speed ever** was 515.3 km/h by the TGV between Courtalain and Tours, France 18 May 1990.

- **The fastest speed on rail** was 9851 km/h by a rocket sled on White Sands Missile Range, New Mexico in Oct 1982.

▼ *TGV Atlantiques often hit 300 km/h on the French part of the Eurostar run from Paris to London.*

Supercars

▲ *The Porsche 911 is one of the classic supercars, first introduced in 1964, but still looking sleek many years later.*

- **The Mercedes Benz 300SL** of 1952 was one of the first supercars of the post-war years, famous for its stylish flip-up 'Gullwing' doors.

- **The Jaguar E-type** was the star car of the early 1960s with incredibly sleek lines and 250 km/h performance.

- **The Ford Mustang** was one of the first young and lively 'pony' cars, introduced in 1964.

- **The Aston Martin DB6** was the classic supercar of the late 1960s, driven by film spy James Bond.

- **The Porsche 911 turbo** was the fastest accelerating production car for almost 20 years after its launch in 1975, scorching from 0 to 100 km/h in 5.4 seconds.

- **The Lamborghini** Countach was the fastest supercar of the 1970s and 1980s, with a top speed of 295 km/h.

- **The McLaren F1** can accelerate from 0-160 km/h in less time than it takes to read this sentence.

- **The McLaren F1** can go from 0 to 160 km/h and back to 0 again in under 20 seconds.

- **A tuned version** of the Chevrolet Corvette, the Callaway SledgeHammer, can hit over 400 km/h.

> ★ **STAR FACT** ★
> The Ford GT-90 has a top speed of 378 km/h
> and zooms from 0-100 kmh in 3.2 seconds.

A sailor's life

- **Sailors** used to be called 'tars' after the tarpaulins used for making sails. Tarpaulin is canvas and tar.

- **A sailor's duties** included climbing masts, rigging sails, taking turns on watch and swabbing (cleaning) decks.

- **The eight-hour** watches were timed with an hour glass.

- **Storing food** on ships in the days before canning was a problem. Sailors survived on biscuits and dried meat.

- **Hard tack** was hard biscuits that kept for years – but often became infested with maggots. The maggots were picked out before the biscuit was eaten.

- **Every sailor** had a daily ration of water of just over a litre since all water had to be carried on board.

- **Sailors** often made clothes from spare materials such as sail canvas.

- **Sailors who offended** against discipline might be flogged with a 'cat-o'-nine-tails' – a whip with nine lengths of knotted cord.

- **Sailors slept** in any corner they could find, often next to the cannon or in the darkness below decks.

- **Sailors' lavatories** on old ships were simply holes overhanging the sea called 'jardines', from the French for garden.

▼ *The introduction of the hammock in the 18th century made sleeping much less uncomfortable for sailors.*

Theatres

▶ There were several Wooden O theatres like this in Elizabethan London in the late 16th century – including the famous Globe where Shakespeare first staged his plays.

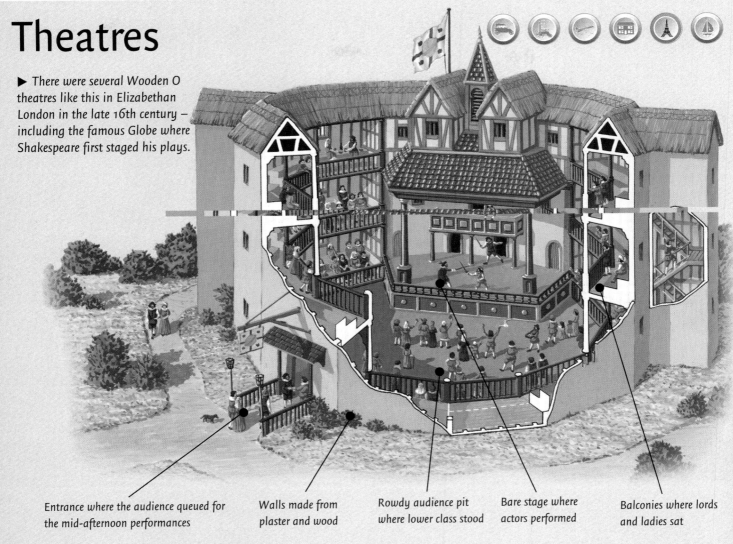

Entrance where the audience queued for the mid-afternoon performances

Walls made from plaster and wood

Rowdy audience pit where lower class stood

Bare stage where actors performed

Balconies where lords and ladies sat

- **In the days of Ancient Greece and Rome** thousands of people went to see dramas in huge, stone-built open-air theatres like sport arenas.

- **The Wooden Os** which appeared in England around 1570 were the first real theatres since Roman times. They were round wooden buildings with the stage and audience arena in the middle, open to the sky.

- **The 17th century** was the 'Golden Age' of drama in Spain, and people crammed into open courtyard theatres called corrales to see plays by great writers, such as Caldéron de la Barca and Lope de Vega.

- **In the 17th century** theatre moved indoors for the first time. Performances were by candlelight in large halls with rows of balconies round the edge.

- **In early theatres** the actors mostly performed on a bare stage. Towards the end of the 18th century, cut-out boards with realistic scenes painted on them were slid in from the wings (the sides of the stage).

- **In the early 19th century** theatres grew much bigger to cater for the populations of the new industrial cities, often holding thousands. Huge stages were illuminated by 'limelight' – brightly burning pellets of calcium.

- **Most of the grand old theatres** we can visit today date from the mid 19th century.

- **The big theatres** of the early 19th century had huge backstage areas with lots of room for 'flying in' scenery, and also spring loaded trapdoors and other ingenious mechanisms for creating giant special effects, like exploding volcanoes and runaway trains.

- **Towards the end of the 19th century** dedicated amateurs, fed up with special effects, performed in smaller spaces, such as clubs, barns and private houses. These developed into intimate 'studio' theatres where actors could stage subtle, realistic plays.

- **Many modern theatres** are equipped with very bright, computer-controlled lighting, sound systems and motorized stage machinery.

Submarines

- **The first workable submarine** was a rowing boat covered with waterproofed skins, built by Dutch scientist Cornelius Van Drebbel in 1620.

- **In 1776** David Bushnell's one-man sub USS *Turtle* attacked British ships in America's War of Independence.

- **Petrol engines** and electric batteries were combined to make the first successful subs in the 1890s.

- **Powerful,** less fumy diesel engines took over from 1908.

- **In 1954** the US launched the *Nautilus*, the first nuclear power sub. Now all subs are nuclear -powered.

- **U-boats** were German subs that attack Allied convoys of ships in World War I and II.

- **In a modern sub** a strong hull of steel or titanium stops it from being crushed by the pressure of water.

> ★ STAR FACT ★
> The tower on top of a submarine is called the sail or conning tower.

▶ To gain weight for a dive, submarines fill their 'ballast' tanks with water. To surface, they empty the tanks.

- **Most subs** are designed for war and have missiles called torpedoes to fire at enemy ships.

- **Attack** subs have guided missiles for attacking ships. Ballistic subs have missiles with nuclear warheads for firing at targets on land. The Russian navy has subs that can strike targets 8000 km away.

Model T Ford

- **In 1905** most cars were 'coach-built' which meant they were individually built by hand. This made them the costly toys of the rich only.

- **The dream of Detroit farmboy** Henry Ford was to make 'a motor car for the great multitude – a car so low in price that no man making a good fortune will be unable to own one.'

◀ The tough, cheap and reliable Model T – the first mass-produced car – put America on the road for the first time and earned the affection of two generations of American families.

- **Ford's solution** was to make a car, which he called the Model T, by the same mass production techniques also used to make guns at the time.

- **Mass production** meant using huge teams of men, each adding a small part to the car as it moved along the production line.

- **Body-panels** for the Model T were stamped out by machines, not hammered by hand as with earlier cars.

- **In 1908** when Ford launched the Model T, fewer than 200,000 people in the USA owned cars.

- **In 1913** 250,000 people owned Model Ts alone.

- **By 1930** 15 million Model Ts had been sold.

- **One of the keys to the T** was standardizations. Ford said early Ts were available in 'any colour they like so long as it's black.' Later models came in other colours.

- **The T's fragile-looking chassis** earned it the nickname 'Tin Lizzie', but it was made of tough vanadium steel.

Bridges of wood and stone

- **The first bridges** were probably logs and vines slung across rivers to help people across.
- **The oldest known bridge** was an arch bridge built in Babylon about 2200 BC.
- **Clapper bridges** are ancient bridges in which large stone slabs rest on piers (supports) of stone.
- **There are clapper bridges** in both Devon in England and Fujian in China.
- **The first brick bridges** were built by the Romans – like the Alcántara bridge over the Tagus River in Spain which was built around 100 AD.
- **Long bridges** could be made with a series of arches linked together. Each arch is called a span.
- **Roman arches** were semi-circular, so each span was short; Chinese arches were flatter so they could span greater distances.
- **The Anji bridge** at Zhao Xian in China was built in 610 AD and it is still in use today.

▲ *The Ponte Vecchio in Florence, built in 1345, is one of the oldest flattened arch bridges in Europe.*

- **Flattened arched bridges** were first built in Europe in the 14th century. Now they are the norm.
- **London Bridge** was dismantled and reconstructed stone by stone in Arizona as a tourist attraction in 1971.

Autogiros and microlights

- **In the 1400s** many European children played with flying toys kept aloft by whirling blades.
- **The autogiro** was invented by the Spanish inventor Juan de la Cierva in 1923.
- **An autogiro** is lifted, not by wings, but by turning rotor blades.
- **A helicopter uses** a powerful motor to turn the rotors; an autogiro's rotors are turned round by the pressure of air as the plane flies forward.
- **The autogiro** is pulled forward by propeller blades on the front like an ordinary small plane.
- **The autogiro** can fly at up to 225 km/h, but cannot hover like a helicopter.
- **In the USA and Australia** microlights are called ultralights. They are small, very light aircraft.
- **The first microlight** was a hang-glider with a chainsaw motor, built by American hang-glider pioneer Bill Bennett in 1973.
- **Some microlights** have flexible fabric wings like hang-gliders.
- **Some microlights** have fixed wings with control flaps to steer them in flight.

◀ *For a while in the 1930s, many people believed autogiros would be the Model T Fords of the air – aircraft for everyone.*

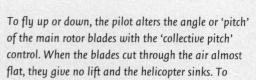

Helicopters

Without a tail rotor, the helicopter would spin round the opposite way to the main rotors. This is called torque reaction. The tail rotor also acts as a rudder to swing the tail left or right

To fly up or down, the pilot alters the angle or 'pitch' of the main rotor blades with the 'collective pitch' control. When the blades cut through the air almost flat, they give no lift and the helicopter sinks. To climb, the pilot steepens the pitch to increase lift

Tail rotor drive shaft

To fly forwards or back, or for a banked turn, the pilot tilts the whole rotor with the 'cyclic pitch' control

The angle of the blades is changed via rods linked to a sliding collar round the rotor shaft, called the swashplate

Rockets

Engine

Stabilizers

▶ A helicopter's rotor blades are really long, thin wings. The engine whirls them round so that they cut through the air and provide lift just like conventional wings (see taking off). But they are also like huge propellers, hauling the helicopter up just as a propeller pulls a plane.

● **Toy helicopters** have been around for centuries, and those made by air pioneer Sir George Cayley in the early 19th century are the most famous.

● **On 13 November 1907** a primitive helicopter with two sets of rotors lifted French mechanic Paul Cornu off the ground for 20 seconds.

● **The problem** with pioneer helicopters was control. The key was to vary the pitch of the rotor blades.

● **In 1937** German designer Heinrich Focke built an aircraft with two huge variable pitch rotors instead of wings and achieved a controlled hover. Months later, German Anton Flettner built the first true helicopter.

● **Focke and Flettner's** machines had two rotors turning in opposite directions to prevent torque reaction. In 1939, Russian born American Igor Sikorsky solved the problem by adding a tail rotor.

● **The Jesus nut** that holds the main rotor to the shaft got its name because pilots said, "Oh Jesus, if that nut comes off…".

● **The biggest helicopter** was the Russian Mil Mi-12 Homer of 1968 which could lift 40,204 kg up to 2255 m.

● **The fastest helicopter** is the Westland Lynx, which flew at 402 km/h on 6 August 1986.

● **The Boeing/Sikorsky RAH-66** Comanche unveiled in 1999 is the first helicopter using stealth technology (see warplanes). It is made of carbon-fibre and other composite materials, and the rotor hub is hidden.

▲ The Vietnam war saw the rise of heavily armed helicopter gunships designed to hit targets such as tanks.

> **! NEWS FLASH !**
> The Bell Quad Tiltrotor has wings like a plane for fast flying. The propellers on the end of each wing tilt up in 20 seconds for vertical lift off. It could evacuate 100 people from danger in minutes.

Dams

▲ The gates of the Thames Barrier in London are designed to shut in times of very high tides to stop seawater flooding the city.

- **The earliest known dam** is a 15 m high brick dam on the Nile River at Kosheish in Egypt, built around 2900 BC.

- **Two Ancient Roman** brick dams, at Proserpina and Cornalbo in Southwest Spain, are still in use today.

- **In China** a stone dam was built on the Gukow in 240 BC.

- **Masonry dams** today are usually built of concrete blocks as gravity dams, arch dams or buttress dams.

- **Gravity dams** are very thick dams relying entirely on a huge weight of concrete to hold the water. They are very strong, but costly to build.

- **Arch dams** are built in narrow canyons and curve upstream. Buttress dams are thin dams strengthened by supports called buttresses on the downstream side.

- **Embankment dams** or fill dams are simple dams built from piles of earth, stones, gravel or clay.

- **The Aswan dam** on the Nile in Egypt has created Lake Nasser – one of the world's biggest artificial lakes.

- **The world's highest dams** are fill dams in Tajikistan – the Rogun (335 m) and Nurek (300 m). The highest gravity dam is the 285 m Grand Dixence in Switzerland. The highest arch dam is the 272 m Inguri in Georgia.

> ★ STAR FACT ★
> The world's biggest dam is the 255 m high, 112 million cu metre Kambaratinsk Dam in Russia.

Warships

- **21st-century navies** have five main classes of surface warship in descending size order: aircraft carriers, landing craft, cruisers, destroyers and frigates.

- **The biggest** warships are 332.9 m-long US aircraft carriers Nimitz, Dwight D. Eisenhower, Carl Vinson, Abraham Lincoln, John C. Stennis, George Washington and Harry S. Truman.

- **In World War II** cruisers had big guns and often acted independently; destroyers protected the main fleet; and frigates protected slower ships against submarines.

- **The distinction between** classes is now blurred. Cruisers are rare, and many small warships carry helicopters or even VTOL planes (see warplanes).

▶ A destroyer armed with a rotating gun for fleet escort.

- **The Russian** Moskva class are a cross between cruisers and aircraft carriers with rear flight decks for helicopters.

- **The British** Invincible class are small aircraft carriers able to carry six Harrier jump jets and twelve helicopters.

- **Warships** have largely replaced guns with missiles.

- **Warships** have both short-range missiles to fire against missiles and long-range supersonic missiles to fire against ships up to 500 km away.

- **The US Ticonderoga** ships are small cruisers built in the 1970s. They are powered by gas-turbine engines and armed with Tomahawk nuclear or conventional missiles.

- **The nuclear-powered** Russian Kirov class begun in 1973 are among the few big cruisers, over 22,000 tonnes.

Steam locomotives

- **Steam locomotives** get their power by burning coal in a firebox. This heats up water in a boiler, making steam. The steam drives a piston to and fro and the piston turns the wheels via connecting rods and cranks.

- **It takes about three hours** for the crew to get up enough steam to get a locomotive moving.

- **Coal and water** are often stored in a wagon called a tender, towed behind the locomotive.

- **A tender** holds 10 tonnes coal and 30,000 litres water.

- **Loco classes** are described by their wheel layout.

- **A 4-6-2** has four small leading 'bogie' wheels, six big driving wheels and two small trailing wheels. The small bogie wheels carry much of the weight.

- **The greatest 19th-century** loco designer was James Nasmyth.

- **In the American Civil War** (1861-65) *The General* was recaptured by Confederates after a chase in another loco.

- **The Flying Scotsman** was a famous loco designed by Sir Nigel Gresley (1876-1941). It pulled trains non-stop 630 km from London to Edinburgh in under six hours.

> ★ STAR FACT ★
> The first loco to hit 100 mph (160 km/h) was the City of Truro in 1895.

▼ *An American locomotive of the 1890s.*

Dhows

- **Dhows** are wooden Arab boats that have been used in the Mediterranean and Indian Ocean for thousands of years.

- **Most dhows** are now made of teak (or Glass Reinforced Plastic). In the past, mango wood was common.

- **In the past** dhows were made, not by nailing planks of wood together, but by sewing them with coconut fibre.

- **Dhow builders** rarely work from plans. They judge entirely by eye and experience.

- **Dhows traditionally** had lateen sails – triangular sails in line with the boat. This allows them to sail almost into the wind.

- **Although many dhows** are now motorized, they usually have a tall mast for unloading.

- **There are several kinds** of dhow, including the shu'ai, the boum, the sambuq, the jelbut and the ghanjah.

- **The jelbut** is thought to get its name from the English 'jolly boat' – the little boats visiting British East Indiamen sent ashore. Jelbuts were often used for pearl fishing. Racing jelbuts have a tree-trunk-like bowsprit almost as long as the hull to carry extra sails.

- **Sambuqs and ghanjahs** are thought to get their square sterns from 17th-century Portuguese galleons.

- **Boums** can be up to 400 tonnes and 50 m long.

▼ *Dhows like this are still built today in Lamu in Africa, much as they were 1000 or more years ago.*

Towns and cities

▲ *The city of Florence is one of the most beautiful in the world. Many buildings dating from the time when European towns began to flourish in the 15th century – especially in Italy.*

- **The first cities** may date from the time when hunters and gatherers settled down to farm 10,000 years ago.

- **The city of Jericho** has been settled for over 10,000 years.

- **One of the oldest** known cities is Catal Hüyük in Anatolia in Turkey, dating from earlier than 6000 BC.

- **The first Chinese cities** developed around 1600 BC.

- **The cities of Ancient Greece** and Rome were often carefully laid out, with major public buildings. The first known town planner was the Ancient Greek Hippodamus, who planned the city of Miletus.

- **In the 1200s** Paris was by far the biggest city in Europe with a population of 150,000.

- **In the 1200s** 320,000 people lived in Hangzhou in China; Guangzhou (Canton) was home to 250,000.

- **In the 1400s** towns like Genoa, Bruges and Antwerp grew as trading centres. Medieval towns were rarely planned.

- **In the 1800s** huge industrial cities based on factories grew rapidly. Chicago's population jumped from 4000 in 1840 to 1 million in 1890.

- **The fastest growing** cities are places like Sao Paulo in Brazil where many poor people live in quickly erected shanty towns. The world's biggest city is Mexico City.

Luxury cars

- **The first Rolls-Royce** was made by Charles Rolls and Henry Royce. It was known as the 'best car in the world'.

- **The Rolls-Royce Silver Ghost** got its name for its ghost-like quietness and shiny aluminium body.

- **Today** each Rolls-Royce takes three months to build.

- **The winged girl** statuette on the bonnet of a Rolls-Royce is called 'Spirit of Ecstasy' and dates from 1911.

- **In the 1930s** Ettore Bugatti set out to build the best car ever with the Bugatti Royale. Only six were built, and they are now the world's most valuable cars.

- **The name Royale** comes from the King of Spain, Alfonso XIII, who was to buy the first model.

- **In 1925** drinksmaker André Dubonnet had a special version of the Hispano-Suiza H6B built with a body made entirely of tulipwood.

- **In the 1930s** American carmakers like Cord, Auburn and Packard made magnificent cars that Hollywood stars posed beside and Chicago gangsters drove.

- **Every 1934** Auburn speedster was sold with a plaque certifying it had been driven at over 100 mph (160 km/h) by racing driver Ab Jenkins.

- **The Mercedes in** Mercedes-Benz was the name of the daughter of Emil Jellinek, who sold the cars in the USA.

▶ *Bentley is one of the great names in luxury cars. But founder Owen Bentley believed in racing as advertisement. The Bentley name gained lasting fame with success in the Le Mans 24-hour races of the 1920s. Seen here is a 1950s rally entry.*

Towers

- **Romans, Byzantines** and medieval Europeans built towers in city walls and beside gates to give platforms for raining missiles on enemies.

- **Early churches** had square towers as landmarks to be seen from afar. From the 1100s, European cathedrals had towers called steeples, topped by a pointed spire.

- **Spires began** as pyramids on a tower, but got taller and were tapered to blend into the tower to make a steeple.

- **In the 17th and 18th centuries** church spires became simple and elegant, as in Park Street Church, Boston, USA.

- **The tallest unsupported tower** is Toronto's 553 m-high CN tower.

- **The tallest** tower supported by cables is the 629 m TV broadcast tower near Fargo and Blanchard in the USA.

- **The Tower of Babel** was a legendary tower built in ancient Babylon in the Near East. The Bible says God didn't want this high tower built, so he made the builders speak different languages to confuse them.

- **The Pharos** was a 135 m lighthouse built around 283 BC to guide ships into the harbour at Alexandria in Egypt.

- **The Tower of Winds,** or Horologium, was built in Athens around 100 BC to hold a sundial, weather vane and water clock.

- **Big Ben** is the bell in St Stephen's Tower in London's Houses of Parliament. The tower once had a cell where 'rioters' like suffragette Emmeline Pankhurst were held.

◀ The Leaning Tower of Pisa in Italy is a 55 m high belltower or 'campanile'. Building began in 1173, and it started to lean as workers built the third storey. It is now 4.4 m out of true.

The first houses

- **It was once thought** that all prehistoric humans lived in caves, but most caves were for religious rituals. The earliest houses were made of materials like wood, leaves, grass and mud – so traces have long since vanished.

- **Stone Age people** probably lived in round huts, with walls of wooden posts and thatches of reed.

- **In Britain** post holes and hearths have been found dating back 10,000 years in places like London's Hampstead Heath and Broom Hill in Hampshire.

- **Early mudbrick** houses dating from at least 9500 years ago are found in Anatolia in the Middle East.

- **Low walls** of big stones were the base for thatched roofs 6500 years ago at places like Carn Brea in Cornwall.

- **Two-storey houses** in Mohenjo Daro in Pakistan from 5000 years ago were built from sun-dried mud bricks and had courtyards, doors, windows and bathrooms.

- **Big Ancient Egyptian** houses had three main areas – a reception room for business, a hall in the centre for guests and private quarters at the back for family.

- **Tomb models** show what Egyptian homes were like.

- **Big Roman country houses** were called villas; a town house was called a domus.

- **Villas** had tiled roofs, verandahs, central heating, marbled floors, lavishly decorated walls and much more.

◀ Some home-building styles have changed little in thousands of years. These houses are in French Polynesia.

Record breaking cars

- **The first car speed record** was achieved by an electric Jentaud car in 1898 at Acheres near Paris. Driven by the Comte de Chasseloup-Laubat, the car hit 63.14 km/h. Camille Jenatzy vied with de Chasseloup-Laubat for the record, raising it to 105.85 km/h in his car *Jamais Contente* in 1899.

- **Daytona Beach** in Florida became a venue for speed trials in 1903. Alexander Winton reached 111 km/h in one of his Bullet cars.

- **The biggest engine** ever to be raced in a Grand Prix was the 19,891cc V4 of American Walter Christie, which he entered in the 1907 French Grand Prix. Later the Christie was the first front-wheel drive car to win a major race – a 400 km race on Daytona Beach.

- **The record** for the outer circuit at Brooklands in England was set in 1935 by John Cobb in a Napier-Railton at 1 min 0.41 sec (23,079 km/h) and never beaten.

▶ Donald Campbell took on the record-breaking mantle of his father —and the Bluebird name for his car. On 17 July 1964, Campbell's Bluebird hit a world record 690.909 km/h on the salt flats at Lake Eyre in South Australia. At one moment he hit over 716 km/h — faster than any wheel-driven car has ever been.

- **In 1911** the governing body for the land speed record said that cars had to make two runs in opposite directions over a 1 km course to get the record.

- **In 1924** Sir Malcolm Campbell broke the Land Speed Record for the first of many times in a Sunbeam at 235.17 km/h. In 1925 he hit 150 mph (242 km/h) for the first time. But his most famous record-breaking runs were in the 1930s in his own *Bluebirds*.

- **In 1947** John Cobb drove with tremendous skill to reach 634.27 km/h in his Railton-Mobil. This stood as the record for 17 years.

- **In 1964** the rules were changed to allow jet and rocket-propelled cars to challenge for the Land Speed Record. The next year Craig Breedlove drove his three-wheeler jet *Spirit of America* to over 500 mph (846.78 km/h).

- **In 1970** Gary Gabelich set the record over 1000 km/h with his rocket-powered *Blue Flame*. This record wasn't beaten until Richard Noble roared to 1019.37 km/h in his Rolls-Royce jet-powered *Thrust 2* in 1983.

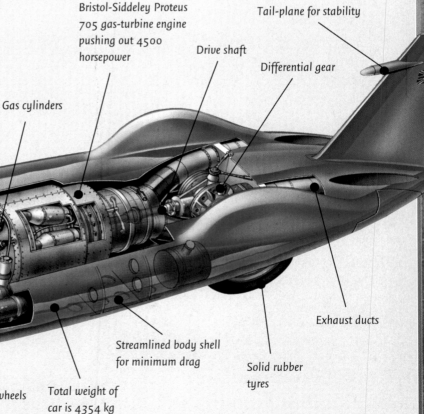

Bristol-Siddeley Proteus 705 gas-turbine engine pushing out 4500 horsepower

Tail-plane for stability

Drive shaft

Differential gear

Gas cylinders

Air intake

Tiny windscreen providing driver's only view

Driver's cockpit

Lightweight aluminium wheels

Total weight of car is 4354 kg

Streamlined body shell for minimum drag

Solid rubber tyres

Exhaust ducts

Ocean liners

▲ One of the great transatlantic liners of the 1930s, the 'Queen Mary' (sister-ship of the 'Queen Elizabeth') is now a hotel at Long Beach California.

- **The great age** of ocean liners lasted from the early 1900s to the 1950s.

- **Ocean liners** were huge boats, often with luxurious cabins, bars, games rooms and swimming pools.

- **The main route** was across the Atlantic. From 1833 liners competed for the Blue Riband title for fastest crossing.

- **Great Blue Riband** contenders included Brunel's *Great Western* in the 1830s, the *Mauretania* which held it from 1907 to 1929, and the French *Normandie* of the 1930s.

- **The last** ocean liner to hold the Blue Riband was the *United States* in the 1960s.

- **The famous Cunard line** was set up by Nova Scotia Quaker Samuel Cunard with George Burns and David MacIver in 1839.

- **The *Titanic*** was the largest ship ever built at 46,329 tonnes when launched in 1912 – but it sank on its maiden (first) voyage.

- **The sinking** of the liner *Lusitania* by a German submarine on 7 May 1915 with the loss of 1198 lives spurred the US to join the war against Germany.

- **The *Queen Elizabeth*** launched in 1938 was the largest passenger ship ever built – 314 m long and 83,673 tonnes. It burned and sank during refitting in Hong Kong in 1972.

- **Future liners** may be based on FastShip technology, with a very broad flat hull for high-speed stability.

The Tower of London

- **The Tower of London** is the oldest stone castle in London, started by William the Conqueror after his conquest of England in 1066. The Crown Jewels are kept here.

- **The oldest part of the tower** is the great square keep called the White Tower which dates from 1078.

- **The Tower** later gained two surrounding 'curtain' walls, like other castles.

- **The inner curtain** wall has 13 towers, including the Bloody Tower, Beauchamp Tower and Wakefield Tower.

- **Traitors' Gate** is an arch beside the River Thames. It gets its name from the time when high-ranking traitors were ferried here by boat to be imprisoned in the Tower. They entered this way.

▶ The Tower of London is the most famous medieval castle in the UK.

- **Many people** have been imprisoned here like Princess (later Queen) Elizabeth and her mother Anne Boleyn (though not at the same time) in the 1500s.

- **Little Ease** is a dark 3.3 square metre cell where prisoners could neither stand up nor lie down.

- **Many prisoners were** beheaded, including Sir Thomas More and Sir Walter Raleigh.

- **In 1483** Edward V and his young brother were thought to have been murdered here. Bones which could have been theirs were later found under stairs.

- **Yeoman Warders** are the Tower's special guards. They are nicknamed beefeaters. The name may have come from a fondness for roast beef or the French word *buffetier*.

Greek and Roman building

▲ ▶ *The most famous Greek temple is the Parthenon in Athens, built c.450 BC.*

- **Ancient Greek and Roman** architecture are together known as Classical architecture.
- **The key features** of Classical buildings are solid, elegantly plain geometric shapes including pillars, arches and friezes.

- **The main Greek** building was the temple with its triangular roof of pale stone on rows of tall columns.
- **Greek temples** were designed in mathematical ratios.
- **Roman architect** Vitruvius described three orders (styles) of columns: Doric, Ionic and Corinthian.
- **Each order** has its own character: Doric serious and strong, Ionic graceful and Corinthian rich and festive.
- **The Parthenon** is a Doric temple built by architects Ictinus and Callicrates guided by the sculptor Phidias.
- **Roman buildings** used many arches and vaults (ceilings made from arches joined together).
- **The ruined Baths of Caracalla** in Rome (AD 217) has huge and graceful vaults.
- **Classical** architecture has inspired many imitations over the centuries, including the Palladian style of the 1600s.

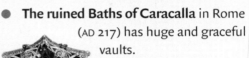

Peoples' cars

- **The first car** for ordinary people was Ford's Model T. Ford built their ten millionth car in 1924 and their 50 millionth in 1959.
- **The USA** began making more than one million cars a year in 1916, of which over a third were Model Ts. No other country made a million cars a year until the UK in 1954.
- **Ford US** introduced weekly payment plans for new cars in 1923. The Nazis later borrowed the idea for the VW Beetle.
- **The Model A Ford** sold one million within 14 months of its launch in December 1927. The Ford Escort of 1980 sold a million in just 11 months.
- **The French Citroën 2CV** was designed to carry 'a farmer in a top hat across a ploughed field without breaking the eggs on the seat beside him'.
- **The VW Beetle** was the brain child of the Nazi dictator Adolf Hitler who wanted a cheap car for all Germans. It was created in 1938 by Ferdinand Porsche.
- **The war interrupted** VW Beetle production before it

had barely begun, but it was resumed after the war.
- **The ten millionth** VW Beetle was built in 1965. Over 22 million have now been sold.
- **The Soviet built Lada,** based on a design by Fiat, was one of the cheapest cars ever built.
- **The first British car** to sell a million was the Morris Minor, between 1949 and 1962.

▼ *The 1959 Mini, designed by Alec Issigonis set a trend in small car design, with the engine mounted across the car driving the front wheels.*

Bridges

▲ London's Tower Bridge was opened in 1894. It has a strong steel frame clothed in stone to support the two opening halves.

- **Rope suspension** bridges have been used for thousands of years. One of the first to use iron chains was the Lan Jin Bridge at Yunnan in China, built AD 65.
- **The first** all-iron bridge was at Coalbrookdale, England. It was designed by Thomas Pritchard and built in 1779 by Abraham Darby.
- **In the early 1800s** Thomas Telford began building superb iron bridges such as Craigellachie over the Spey in Scotland (1814). He built Europe's first iron chain suspension bridge over the Menai Straits in Wales in 1826.
- **Stephenson's** Britannia railway bridge of 1850, also over the Menai Straits, was the first hollow box girder bridge.

▶ Most bridges are now built of concrete and steel. Shown here are some of the main kinds. The longest are normally suspension bridges, usually carrying roads, but Hong Kong's 1377 m-long Tsing Ma (1998) takes both road and rail.

★ **STAR FACT** ★
The Akashi-Kaikyo Bridge (1998), Japan, is the world's longest suspension bridge with a main span of 1991 m and two side spans of 960 m.

- **John Roebling's** Cincinnati suspension bridge was the world's longest bridge when built in 1866 at 322 m. Like all suspension bridges today, it was held up by iron wires, not chains. It was the prototype for his Brooklyn Bridge.
- **The Forth Railway Bridge** in Scotland was the world's first big cantilevered bridge. At 520 m it was also the world's longest bridge when it was built in 1890.
- **In 1940** the Tacoma suspension bridge in Washington USA, was blown down by a moderate wind just months after its completion. The disaster forced engineers to make suspension bridges aerodynamic.
- **Aerodynamic design** played a major part in the design of Turkey's Bosphorus Bridge (1973) and England's Humber Bridge (1983), for a while the longest bridge at 1410 m.
- **In the 1970s** Japanese engineers began to build a bridge that by 2000 gave Japan nine of the world's 20 longest bridges.

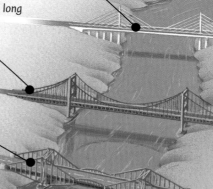

In cable-stayed bridges, the bridge hangs directly from steel cables

In suspension bridges the bridge hangs on steel wires on a cable suspended between tall towers. They are light so can be very long

In cantilevered bridges, each half of the bridge is balanced on a support

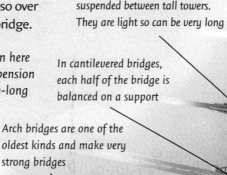

Arch bridges are one of the oldest kinds and make very strong bridges

Bascule or lifting bridges like London's Tower Bridge swing up in the middle to allow tall ships through

Steel or concrete beam bridges are carried on piers. The beam may be a hollow steel girder through which cars and trains can run

The Statue of Liberty

- **New York's Statue of Liberty** stands on Liberty island off the tip of Manhattan.

- **The statue** was dedicated on 28 Oct 1886 by President Cleveland.

- **It was paid** for by the French people to celebrate their friendship with the USA.

- **Sculptor** Frédéric-Auguste Bartholdi began work on the statue in Paris in 1875.

◄ *New York's famous Statue of Liberty, before it was restored in 1986 and the flame covered in gold leaf.*

- **It was built** from 452 copper sheets hammered into shape by hand and mounted on four huge steel supports designed by Eiffel and Viollet-le-Duc.

- **The 225-tonne statue** was shipped to New York in 1885.

- **A pedestal** designed by Richard Hunt and paid for by 121,000 Americans brought it to a total height of 93 m.

- **The statue's** full name is *Liberty Enlightening the World*. The seven spikes in the crown stand for liberty's light shining on the world's seven seas and continents. The tablet in her left arm is America's Declaration of Independence.

- **Emma Lazarus's sonnet** *The New Colossus* on the pedestal ends: "Give me your tired poor, your huddled masses of your teeming shore. Send these, the homeless, tempest-tossed to me. I lift my lamp beside the golden door!"

> ★ **STAR FACT** ★
> The Statue of Liberty's crown houses an observation deck for up to 20 people.

Buses and coaches

- **Horse-drawn stage coaches** were the first regular public coach services between two or more points or 'stages'. They were first used in London in the 1630s.

- **Stage coaches** reached their heyday in the early 1800s when new tarred roads made travel faster. Coaches went from London to Edinburgh in 40 hours.

- **Bus is short** for 'omnibus' which in Latin means 'for all'. The word first came into use in Paris in the 1820s for big coaches carrying lots of people on local journeys.

- **In 1830** Goldsworthy Gurney put a steam engine in a coach to make the first powered bus. It ran four times a day between Cheltenham and Gloucester in England.

- **In the 1850s** British government laws restricted steam road vehicles, so big new cities developed horse buses.

- **In 1895** a petrol-engined bus was built in Germany.

- **In 1904** the London General Omnibus Co. ran the first petrol-engined bus services. They were double-decked.

- **1905:** the first motor buses ran on New York's 5th Avenue.

▲ *Like most early buses, this one from the early 1920s was built by adding a coach body to a lorry base.*

- **1928:** the first transcontinental bus service crossed the USA.

- **From the 1950s** articulated buses were used in European cities. A trailer joined to the bus carries extra passengers.

St Basil's cathedral

- **The cathedral of** St Basil the Blessed in Moscow's Red Square was built from 1554 to 1560.

- **St Basil's** is made up of ten tower churches, the biggest of which is 46 m tall.

- **It began with** eight little wooden churches – each built between 1552 and 1554 after a major Russian victory against the Tartars of Kazan in central Asia.

- **After the final victory**, Ivan the Terrible ordered stone churches to be built in place of the wooden ones.

- **Legend says** each of the onion-shaped domes represents the turban of a defeated Tartar lord.

▲ With its ten famous colourful towers, St Basil's in Moscow one of the most beautiful Christian cathedrals in the world.

- **St Basil's** was designed by two Russians, Posnik and Barma, who may have been one person.

- **Legend says** the Italian builder was blinded afterwards so he could build nothing like it again.

- **Originally** it was known as the Cathedral of the Intercession, but in 1588 a tenth church was added in honour of St Basil and it was known afterwards as St Basil's.

- **St Basil** was the jester (comedian) to a Moscow lord of the time.

- **In 1955**, restorers found the secret of the cathedral's construction embedded in the brickwork – wooden models used as silhouettes to guide the builders.

Hydrofoils

- **Hydrofoils** are boats with hulls that lift up above the water when travelling at high speeds.

- **The hydrofoils** are wings attached to the hull by struts that move underwater like aeroplane wings and lift the boat up.

- **Because only the foils** dip in the water, hydrofoils avoid water resistance, so can travel faster with less power.

- **Surface-piercing hydrofoils** are used in calm inland waters and skim across the surface.

▼ By lifting themselves out of the water and almost flying across the surface, hydrofoils achieve very high speeds.

★ STAR FACT ★
The biggest hydrofoil, built in Seattle, is the 64 m long, 92 km/h *Plainview* navy transport.

- **Full-submerged hydrofoils** dip deep into the water for stability in seagoing boats.

- **The foils** are usually in two sets, bow and stern.

- **The bow and stern foils** are in one of three arrangements. 'Canard' means the stern foil is bigger. 'Airplane' means the bow foil is bigger. 'Tandem' means they are both the same size.

- **The first successful hydrofoil** was built by Italian Enrico Forlanini in 1906.

- **In 1918** Alexander Graham Bell, inventor of the telephone, built a hydrofoil that set a world water speed record at 61.6 knots (114 km/h). The record was not beaten until the American *Fresh 1*, another hydrofoil, set a new record of 84 knots (155 km/h) in 1963.

Balloons

- **Balloons** are bags filled with a light gas or hot air – both so light that the balloon floats in the air.
- **Balloons** designed to carry people into the air are of two types: hot-air balloons and gas balloons filled with hydrogen or helium.
- **Hot-air balloons** have a burner that continually fills the balloon with warm air to keep it afloat.
- **To carry two people** a hot-air balloon must have a bag of about 1700 cubic metres in volume.
- **Balloons** are normally launched at dusk or dawn when the air is quite calm.
- **As the air in the bag cools**, the balloon gradually sinks. To maintain height, the balloonist lights the burner to add warm air again.
- **To descend quickly** the balloonist pulls a cord to let air out through a vent in the top of the bag.
- **The first flight** in a hot air balloon was made in Paris on 15 October 1783 by French scientist Jean de Rozier in a

▶ *Hot air ballooning has become a popular sport since Ed Yost, Tracy Barnes and other Americans began to make the bags from polyester in the 1960s.*

balloon made by the Montgolfier brothers.

- **The first hydrogen gas balloon flight** was made in Paris on 1 December 1783.
- **On 20 March 1999** Swiss Bertran Piccard and British Brian Jones made the first round-the-world hot air balloon flight.

Record breaking flights

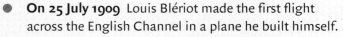

- **On 25 July 1909** Louis Blériot made the first flight across the English Channel in a plane he built himself.
- **In May 1919** Capt A.C. Read and his crew made the first flight across the Atlantic in a Curtiss flying boat.
- **On 14-15 June 1919** John Alcock and Arthur Brown made the first non-stop flight across the Atlantic in an open cockpit Vickers Vimy biplane.

> ★ STAR FACT ★
> In December 1986, *Voyager*, piloted by Rutan and Yeager, flew non-stop round the world in 9 days.

- **In November 1921** brothers Keith and Ross Smith made the first flight from England to Australia.
- **On 4 May 1924** Frenchman Étienne Oehmichen flew his helicopter in a 1 km circle.
- **In 1927** Frenchman Louis Breguet made the first flight across the South Atlantic.
- **On 21 May 1927** American Charles Lindbergh made the first solo flight across the Atlantic in the *Spirit of St Louis*.
- **In July 1933** Wiley Post made the first solo round-the-world flight.
- **The story of Post's epic** flight was told in the book *Round the World in Eight Days*.

◀ *Blériot being greeted near Dover after completing his flight across the English channel.*

Chinese building

- **China** developed its own distinctive style of building over 3000 years ago.
- **Traditionally, large Chinese buildings** were made of wood on a stone base.
- **A distinctive feature** is a large tiled overhanging roof, ending in a graceful upturn. The tiles were glazed blue, green or yellow.
- **The roof** is supported not by the walls but by wooden columns, often carved and painted red and gold.
- **Walls were thin** and simply gave privacy and warmth.
- **Chinese temples** were large wooden halls with elaborate roof beams in the ceiling.
- **Pagodas** are tall tapering towers often linked to Buddhist temples. They have from 3 to 15 storeys.

> ★ **STAR FACT** ★
> Chinese pagodas have eight sides and an uneven number of storeys.

▲ The Forbidden City in Beijing dates mostly from the Ming era from 1368-1644. Only the emperor's household could enter it.

- **In China, pagodas** were believed to bring happiness to the surrounding community. They were made of wood, bricks, tiles or even porcelain and decorated with ivory.
- **Pagodas**, originally from India, developed from Buddhist burial mounds called stupas.

Canoes

- **The first canoes** were scooped out logs.
- **A skin canoe** was made by stretching animal skins over a bent wood frame.
- **A skin canoe** dating from around 4500 BC was found on the Baltic island of Fünen.

◄ ▼ The materials may be different, but today's bright fibreglass canoes are based on the same principles as the ancient bark canoes of native Americans.

- **Some skin canoes** are round like the paracil and the Welsh coracle.
- **Some skin canoes** are long and thin like the Irish curach and the Inuit kayak.
- **Kayaks** have a watertight cover made from sealskin to keep out water in rough conditions.
- **The ancient quffa** of Iraq is a large canoe made of basketwork sealed with tar and dates back to 4000 BC.
- **Native Americans** made canoes from bark. The Algonquins used paper birch, the Iroquois elm.
- **Bark canoes** were the basis for today's sport canoes of fibreglass, plastic and aluminium.
- **Unlike ordinary boats,** canoes are often so light that they can easily be carried overland to avoid waterfalls or to move from one river to another.

Electric trains

- **The first practical** electric trains date from 1879, but they only became widespread in the 1920s.

- **Electric locos** pick up electric current either from a third 'live' rail or from overhead cables.

- **To pick up** power from overhead cables, locos need a spring-loaded frame or pantograph to conduct electricity to the transformer.

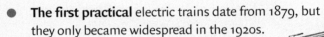

- **Electric trains** are clean and powerful, and can also travel faster than other trains.

- **Older systems** mostly used Direct Current (DC) motors, operating at 1500-3000 volts for overhead cables and 700 volts for live rails.

- **High-speed trains** like France's TGV and Japan's Shinkansen use 'three-phase' Alternating Current (AC) motors operating at 25,000 volts.

◀ *Japan's Shinkansen 'bullet train' was the first of the modern high speed electric trains, regularly operating at speeds of over 400 km/h.*

- **The Paris-London Eurostar** works on 25,000 volt AC overhead cables in France, and 750 volt live rails after it comes out of the Channel Tunnel in England.

- **Magnetic levitation** or maglev trains do not have wheels but glide along supported by electromagnets.

- **In electrodynamic maglevs**, the train rides on repulsing magnets. In electromagnetic maglevs, they hang from attracting magnets.

- **Maglevs** are used now only for short, low-speed trains, but they may one day be the fastest of all. High-speed maglev developments now use 'superconducting' electromagnets which are costly to make. But a new idea is to use long strings of ordinary permanent magnets.

Stations

- **London Bridge** station is the oldest big city terminal, first built from wood in 1836, then of brick in the 1840s.

- **In the 19th century** railway companies competed to make the grandest, most palatial railway stations.

- **The Gare d'Orsay** in Paris was an incredibly luxurious station built in 1900 in what is called the *Beaux Arts* (beautiful arts) style. It is now a museum and gallery.

- **When first built** in 1890, Sirkeci in Istanbul – terminal of the Orient Express – glittered like an oriental palace.

- **Bombay's Victoria** – now Chhatrapati Sivaji – was built in 1888 over a shrine to Mumba Devi, named after the patron goddess of original inhabitants of Bombay.

- **Grand Central Station** in New York cost a staggering 43 million dollars to build in 1914.

- **Liverpool Street Station** in London was built in 1874 on the site of the 13th century Bedlam hospital for lunatics.

- **London's St Pancras** is a stunning Gothic building designed by George Gilbert Scott. It was built 1863-72.

- **Chicago's** North Western is a monument to the Jazz Age in the city, dating from 1911.

- **In April 1917** Lenin returned to Russia and announced the start of the Russian revolution at St Petersburg's famous Finland Railway station.

▼ *The great hall at Grand Central Station in New York is one of the most spectacular railway halls in the world.*

Submersibles

- **Submersibles** are small underwater craft. Some are designed for very deep descents for ocean research. Others are designed for exploring wrecks.

- **One early submersible** was a strong metal ball or bathysphere, lowered by cables from a ship.

- **The bathysphere** was built by Americans William Beebe and Otis Barton who went down 900 m in it off Bermuda on 11 June 1931. The possibility of the cable snapping meant the bathysphere was very dangerous.

- **The bathyscaphe** was a diving craft that could be controlled underwater, unlike the bathysphere. Its strong steel hull meant it could descend 4000 m or more.

- **The first bathyscaphe**, the FNRS 2, was developed by Swiss scientist August Piccard between 1946 and 1948. An improved version, the FNRS 3, descended 4000 m off Senegal on 15 February 1954. The FNRS 3 was further improved to make the record-breaking Trieste.

- **In the 1960s** the Woods Hole Oceanographic Institute in the USA began to develop a smaller, more manoeuvrable submersible, called *Alvin. Alvin* is one of the most famous of all submersibles, making thousands of dives to reveal a huge amount about the ocean depths.

- **ROVs** or Remote Operated Vehicles are small robot submersibles. ROVs are controlled from a ship with video cameras and computer virtual reality systems. ROVs can stay down for days while experts are called in to view results. Using the ROV Argo-Jason, Robert Ballard found the wreck of the liner *Titanic* in 1985.

- ***Deep Flight*** is a revolutionary submersible with wings that can fly underwater like an aeroplane, turning, diving, banking and rolling.

- **A new breed** of small submersibles, like the *Sea Star* and *Deep Rover*, cost about the same as a big car and are designed for sports as well as research.

▶ This is one of the first of the huge range of submersibles that began to appear in the 1960s and 70s. They are now much smaller, neater and more manoeuvrable, but still work in much the same way.

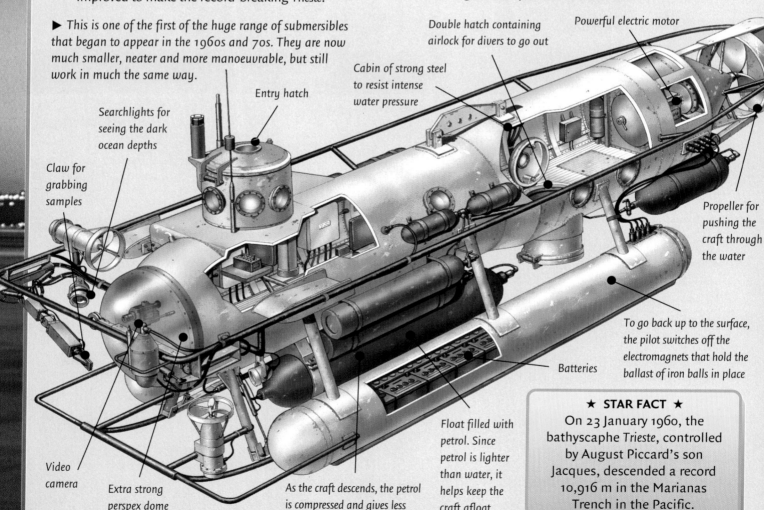

Double hatch containing airlock for divers to go out

Powerful electric motor

Cabin of strong steel to resist intense water pressure

Entry hatch

Searchlights for seeing the dark ocean depths

Claw for grabbing samples

Propeller for pushing the craft through the water

To go back up to the surface, the pilot switches off the electromagnets that hold the ballast of iron balls in place

Batteries

Video camera

Extra strong perspex dome

As the craft descends, the petrol is compressed and gives less buoyancy, speeding the descent

Float filled with petrol. Since petrol is lighter than water, it helps keep the craft afloat

★ STAR FACT ★
On 23 January 1960, the bathyscaphe *Trieste*, controlled by August Piccard's son Jacques, descended a record 10,916 m in the Marianas Trench in the Pacific.

Airports

- **The world's first airport** was built at Croydon near London in 1928. Many early airports, like Berlin's, were social centres attracting thousands of visitors.

- **Before airports,** flying boats would land on water. So airports like New York's La Guardia were set close to water to take flying boats.

- **Over 50 airports** around the world now handle over 10 million passengers a year. 25 of these are in the USA.

- **Six airports** handle over 30 million passengers, including Chicago's O'Hare and Hong Kong's Chep Lap Kok.

- **The world's largest** airport is King Abdul Aziz in Saudi Arabia. It covers 22,464 hectares. The USA's biggest is Dallas. Europe's biggest is Paris's Charles de Gaulle.

- **Hong Kong's** Chep Lap Kok airport, opened in 1998, is one of the world's most modern.

- **Kansai** airport in Japan is built entirely on an artificial island in Osaka Bay so that it can operate 24-hours a day, without disturbing people with noise.

▲ In the 1970s, Boeing 747 jumbo jets needed 4 km runways to take-off, but better performance means they now need less distance.

- **In early airports** terminals for each flight were set in a line as at Kansas and Munich. But as flights increased, this layout meant passengers had a long way to walk.

- **Terminals in the 1970s** were set in extending piers like Amsterdam's Schiphol, or satellites like Los Angeles.

- **New airport terminals** like London's Stansted are set apart and linked by electric cars called 'people-movers'.

Trams and cable cars

- **Trams** are buses that run on rails laid through city streets. They are called streetcars in the USA.

- **Early trams** in the 1830s were pulled by horses. By the 1870s, horse-drawn trams were used in many cities.

- **In 1834** Thomas Davenport, a blacksmith from Vermont, USA built a battery-powered electric tram.

- **In 1860** an American called George Train set up battery-powered electric tram systems in London.

- **In 1873** Andrew Hallidie introduced cable cars in San Francisco. The cars were hauled by a cable running in a slot in the street. A powerhouse pulled the cable at around 14 km/h. Similar systems were built in many cities in the 1880s but were soon replaced by electric trams.

- **In 1888** Frank Sprague demonstrated a tram run from electric overhead cables in Richmond in the USA.

- **In most US trams,** electric current was picked up via a long pole with a small wheel called a shoe that slid along the cable. The pick-up was called a trolley.

Many European trams, however, picked up current via a collapsible frame called a pantograph.

- **In the early 1900s,** electric tram systems were set up in most world cities.

- **In the 1930s** most cities, except in eastern Europe and Russia, replaced trams with buses.

- **In the 1990s** some cities, like Manchester in England, built new tramways, because they are fume-free.

◀ The original Hallidie cable car system dating from 1873 still runs in San Francisco.

The Wright brothers

- **The Wright brothers,** Orville and Wilbur, built the world's first successful plane, the *Flyer.*

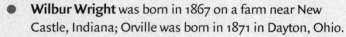

- **On 17 December 1903** the Wright brothers made the first powered, long and controlled airplane flight at Kitty Hawk, USA.

- **Wilbur Wright** was born in 1867 on a farm near New Castle, Indiana; Orville was born in 1871 in Dayton, Ohio.

- **The Wright brothers** began as bicycle makers but became keen on flying after hearing about the death of pioneer glider Otto Lilienthal in 1896.

- **From 1899 to 1903** they worked at Kitty Hawk methodically improving their design and flying skill.

◀ One of the five who witnessed the flight took this picture. But apart from a report in 'Popular Science' the Wrights' success was little known about for five years.

- **Many early planes** took off but lacked control. The key to the Wrights' success was stopping the plane rolling, using wires to 'warp' (twist) the wings to lift one side or the other.

- **The *Flyer*'s** wing warp meant it could not only fly level but make balanced, banked turns (like a bicycle cornering).

- **For the first flight** Orville was at the controls.

- **The historic first flight** lasted 12 seconds, in which the *Flyer* travelled 37 m and landed safely.

- **On 5 October 1905** the Wrights flew 38.9 km in 38 mins.

Bicycles

▲ Bicycles are the cheapest, most reliable form of transport ever, and in the 1920s many tradesmen adapted them for use.

- **The first bicycle** was the 'draisienne' of 1818 of Baron de Drais. The rider scooted his feet on the ground to move.

- **In 1839** Scots blacksmith Kirkpatrick Macmillan invented the first bicycle with pedals and brakes.

- **On Macmillan's 'velocipede'** pedals were linked by rods to cranks on the back wheel. These turned the wheel slowly.

- **In 1861** French father and son Pierre and Ernest Michaux stuck the pedals directly on the front wheel to make the first successful bicycle, nicknamed 'boneshaker'.

- **In 1870** James Starley improved the boneshaker with the Ordinary. A huge front wheel gave high speed with little pedalling.

- **In 1874** H. J. Lawson made the first chain-driven bicycle. This was called a 'safety bicycle' because it was safer than the tall Ordinary.

- **In 1885** Starley's nephew John made the Rover Safety bicycle. Air-filled tyres were added in 1890, and the modern bicycle was born.

- **By 1895** four million Americans were riding bicycles.

- **Today** 50 million people in the USA cycle regularly.

- **More people** in China cycle today than the rest of the world put together.

Great voyages

▶ In 1580, Sir Francis Drake became the first Englishman to sail round the world. En route, he visited what is now California.

- In 330 BC the Greek sailor Pytheas sailed out into the Atlantic through the Straits of Gibraltar and found his way to 'Thule' (Britain).

- Around AD 1000 the Viking Leif Ericsson was the first European to cross the Atlantic to North America.

- In 1405 Chinese admiral Cheng Ho began a series of seven epic voyages around the Indian Ocean in a huge fleet of junks.

- In 1492 Genoese Christopher Columbus crossed the Atlantic from Cadiz in Spain and discovered the New World. His ship was the *Santa Maria*.

> ★ STAR FACT ★
> In 1522, Ferdinand Magellan's ship *Victoria* completed the first round-the-world voyage.

- From 1499-1504 Italian Amerigo Vespucci sailed the Atlantic and realized South America was a continent. America is named after him.

- In 1492 Italian John Cabot crossed the North Atlantic from Bristol in England and found Canada.

- In 1498 Portuguese Vasco da Gama was the first European to reach India by sea, sailing all round Africa.

- In 1642 Dutchman Abel Tasman sailed to what is now Australia and New Zealand. The island of Tasmania is named after him.

- From 1768-79 James Cook explored the South Pacific in his ship the *Endeavour* and landed at Botany Bay.

▶ Like many 15th century explorers, Columbus sailed a 'carrack' with 'lateen' sails for sailing close in to the wind.

Ancient palaces

- **The word palace** comes from the Palatine Hill in Rome, where the emperors of Rome had their palaces.

- **The earliest known** palaces are those built in Thebes in Ancient Egypt by King Thutmose III in the 15th century BC.

- **Egyptian palaces** had a rectangular wall enclosing a maze of small rooms and a courtyard – a pattern later followed by many Asian palaces.

- **The Babylonians** and Persians introduced grand halls in their palaces at Susa and elsewhere.

- **Minoan palaces** on Crete introduced extra storeys.

- **Roman palaces** were the grandest of the ancient world.

- **The Sacred Palace** of Byzantium (now Istanbul) was biggest of all, covering 334,000 square metres.

- **Palaces** in China and Japan, like Beijing's Forbidden City, often had a high wall surrounding small houses for the ruler and his officials.

- **Ancient palaces in the Americas** like the Mayan palace at Uxmal dating from AD 900, are usually simpler.

- **Potala palace** stands high above Lhasa in Tibet. Potala means 'high heavenly realm'. The original, built by King Srong-brtsan-sgam-po in the 7th century was destroyed by the Chinese c.1600. The Dalai Lama rebuilt it in 1645.

◀ The 12th-century Khmer emperors of Cambodia built magnificent palaces in their capital of Angkor, now overgrown by jungle. But most magnificent of all was the temple, Angkor Wat.

Docks and ports

- **Some ports** like Hong Kong and Rio, are based on natural harbours; others are constructed artificially.

- **The Phoenicians** built artificial harbours at Tyre and Sidon in the Lebanon in the 13th-century.

- **The Romans** invented waterproof concrete to build the quays and breakwaters at their port of Ostia.

- **Roman and other ancient ports** had animal and human powered cranes. Saxon London and Viking Dublin had well-built wooden wharves.

- **Traditionally,** big boats would anchor in mid stream and barges called lighters would take the cargo ashore.

- **In the 18th-century** the first enclosed deep water docks were built at London and Liverpool. Here big ships could moor and unload directly onto the wharves.

> ★ **STAR FACT** ★
> The world's busiest ports are Rotterdam, Singapore and Hong Kong.

- **In the 20th century,** ports specializing in particular cargoes, such as oil terminals, grew.

- **Since the 1950s** there has been a huge growth in container ports. Containers are big standardized metal crates that can be loaded on and off quickly.

- **The world's main container ports** are Hong Kong, Singapore, Rotterdam, Hamburg and New York.

▲ The port of Balstad in Norway, a perfect natural harbour.

Modern architecture

▶ The old and new in Hong Kong: the ultra modern Hong Kong-Shanghai bank dwarfs a 19th-century classical building.

- **In the 1920s** many architects rejected old styles to experiment with simple shapes in materials like glass, steel and concrete.

- **The International Style** was pioneered by architect Le Corbusier who built houses in smooth geometric shapes like boxes.

- **The Bauhaus** school in Germany believed buildings should look like the job they were meant to do.

- **Walter Gropius** and Mies van de Rohe took Bauhaus ideas to the US and developed sleek, glass-walled, steel-framed skyscrapers like New York's Seagram Building.

- **American Frank Lloyd Wright** (1869-1959) was famous both for his low, prairie-style bungalows 'growing' from their site and his airy and elegant geometric buildings.

- **In the 1950s** architects like Kenzo Tange of Japan reacted against the 'blandness' of the International Style, introducing a rough concrete look called Brutalism.

- **In the 1960s** many critics reacted against the damage done by modern architecture to historic cities.

- **Post-modernists** were united in rejecting modern architecture, often reviving historical styles. American Robert Venturi added traditional decoration.

- **Richard Rogers'** Pompadou centre in Paris (1977) was a humorous joke on the Bauhaus idea, exposing the 'bones' of the building.

- **With shiny metal** and varied shapes the Guggenheim gallery in Bilbao in Spain is a new masterpiece.

Pyramids

- **Pyramids** are huge ancient monuments with a square base and triangular sides coming together in a point at the top. Remarkably, they are found not only in Egypt, but also in Greece, Cyprus, Italy, India, Thailand, Mexico, South America and various Pacific islands.

- **The earliest tombs** of the pharaohs were mud chambers called mastabas. But c.2686 BC, the wise man Imhotep had a pyramid-shaped tomb built for Zoser at Saqqara. This rose in steps and so is called the Step Pyramid.

- **In 2600BC** the Egyptians filled in the steps on a pyramid with stones to make the first smooth pyramid at Medum.

- **The ruins of 80 pyramids** stand near the Nile in Egypt, built over 1000 years. Each has an elaborate system of hidden passageways and chambers to stop robbers getting at the tomb of the pharaoh inside.

- **The largest pyramid** is the Great Pyramid at Giza, built for the Pharaoh Khufu c.2551-2472 BC. This was once 147 m high, but is now 140 m, since some upper stones have been lost. Khufu is known as Cheops in Greek.

> ★ **STAR FACT** ★
> The Great Pyramid at Giza was made of
> 2.3 million 2.5 tonne blocks of limestone.

▶ The strange half lion, half pharaoh statue called the Sphinx stands near the pyramid of pharaoh Khafre at Giza.

- **Later,** the Egyptians abandoned pyramids and cut tombs into rock, as in the Valley of Kings at Thebes. The tomb of Tutankhamun was one of these.

- **The Moche Indians** of Peru built large brick pyramids, including the Pyramid of the Sun near Trujillo.

- **Another Pyramid of the Sun** is on the site of the ancient Aztec capital of Teotihuacan, 50 km from Mexico City. Built around 2000 years ago it is exactly the same size (230 m square) as the Great Pyramid at Giza.

- **The American pyramids** were not tombs like those in Egypt, but temples, so they have no hidden chambers inside. They are built not from blocks of cut stone but from millions of basketloads of volcanic ash and gravel.

▶ No-one knows just how the Egyptians managed to build such huge structures as the pyramids. It is thought that over 100,000 people worked on each big pyramid. Each big block was probably cut from the quarry, dragged on rollers to the River Nile and carried on barge to the building site. The stones were then dragged into place up earth ramps.

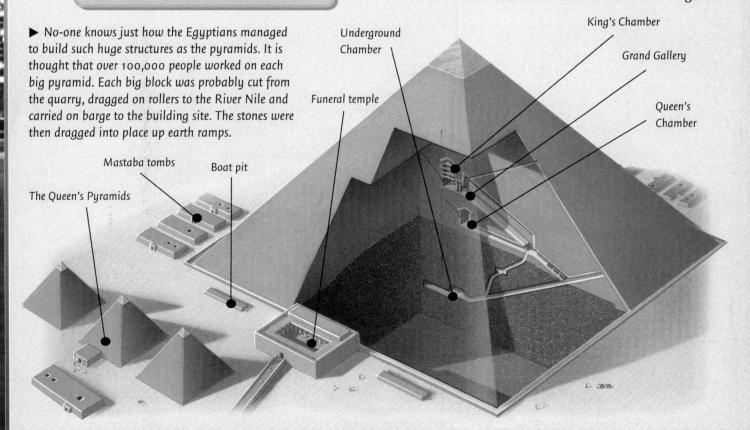

King's Chamber

Grand Gallery

Queen's Chamber

Underground Chamber

Funeral temple

Mastaba tombs

Boat pit

The Queen's Pyramids

Moving on snow

▲ *Cutters are still used in the north eastern USA as a charming way of getting around in the winter snows.*

- **Vehicles** designed to move on snow are supported on flat boards called runners, skids or skis. These slide over snow easily and spread the vehicle's weight over a larger area.

- **Sleds or sleighs** drawn by horses or dogs may have been the first vehicles ever used by humans.

- **Cutters** are light, graceful horse-drawn sleighs first introduced in the USA about 1800.

- **Troikas** are Russian sleighs (or carriages) drawn by three horses. The middle horse is supposed to trot while the outer horses gallop in particular ways.

- **Native Americans** had toboggans made of poles tied with leather thongs.

- **Snowmobiles** are vehicles with two skis at the front and a motor-driven track roll at the back. Racing snowmobiles can reach over 160 km/h.

- **Snowmobiles** are steered by handlebars that control the skis and by the shifting of the driver's weight.

- **The first propeller-driven** snowmobile was built in the 1920s. Tracked snowmobiles were developed in 1959 by Canadian Joseph Bombardier.

- **In the 1960s,** carved wooden skis dating back 8000 years were found in a bog at Vis near the Urals in Russia.

- **The earliest skates** were made from small bones.

Rescue boats

- **Several designs** for 'unsinkable' lifeboats were tried in 18th-century France and England.

- **After many drowned** when the ship *Adventure* went down in 1789, a competition for a lifeboat was set up.

- **The competition winner** was Newcastle boatbuilder Henry Greathead. His boat the *Original* would right itself when capsized and still float even when nearly filled with water.

- **The first land-based** powered lifeboat was steam powered and launched in 1890.

- **Around 1907** diesel-power lifeboats were introduced. Without the need for oars, most

of the boat could be covered in and so made even more unsinkable.

- **Modern land-based lifeboats** are either steel-hulled or made of a double-skin of timber.

- **The hull** usually contains a large number of sealed air containers so it is almost impossible for the boat to sink.

- **Some lifeboats** are afloat all the time. Others are kept ashore in a lifeboat house ready for emergencies.

- **Lifeboats** kept ashore are launched down a slipway into the sea or wheeled into the sea on a cradle until they float.

- **As leisure sailing** and bathing has increased, so there has been an increase in the number of inshore rescue boats. These are small rubber or fibreglass boats, kept afloat by an inflatable tube of toughened rubber.

▶ *This is a typical inflatable inshore rescue boat. Boats like these are bonded together by heat and glue, and are incredibly tough.*

Famous trains

- **Pullman coaches** became a byword for luxury in the USA in the 1860s.

- **After a trip to the USA,** young Belgian George Nagelmackers set up the Wagon-Lits company to do in Europe as Pullman had in the USA.

- **The most famous** Wagon-Lits was the luxurious *Orient-Express* from Paris to Istanbul, which started in 1883.

- **Among the many famous** travellers on the *Orient Express* were the female spy Mata Hari, the fictional spy James Bond and 40s film star Marlene Dietrich.

- **Another famous** Wagon-Lits train is the Trans-Siberian express, set up in 1898.

- **The Trans-Siberian** takes eight days to go right across Russia and Siberia from Moscow to Vladivostok.

> ★ STAR FACT ★
> At 9438 km, the Trans-Siberian Express is the world's longest train journey.

▲ The Orient Express gained the glamour of intrigue as well as luxury after Agatha Christie wrote 'Murder on the Orient Express'.

- **The *Flying Scotsman*** was famed for its fast, non-stop runs from London to Edinburgh in the 1920s.

- **The *Golden Arrow*** ran from London to Paris.

- **The *Indian Pacific*** in Australia runs on the world's longest straight track, 478 km across the Nullarbor Plain.

Sailing and tacking

- **Sailing ships** can sail into the wind because the wind does not so much push the sail as suck it.

- **The sail** is always angled so that the wind blows across it, allowing it to act like the wing of an aircraft.

- **As the wind blows** over the curve of the sail, it speeds up and its pressure drops in front of the sail. The extra pressure of the other side drives the boat forward.

- **The sail** works as long as the wind is blowing straight across it, so to go in a particular direction, the sailor simply changes the angle of the sail.

- **The boat's keel,** or centreboard, stops it slipping sideways.

▲ With their sails in line with the hull, these yachts are sailing close to the wind.

- **The boat can sail** with the wind behind it, to the side of it or even slightly ahead of it.

- **To sail directly** or nearly directly into the wind, a boat has to 'tack'. This means zig-zagging to and fro across the path of the wind so wind is always blowing across the sails.

- **At the end** of each tack, the sail 'comes about' (changes course). The tiller (steering arm) is turned to point the boat on the new course and the sail is allowed to 'gybe' (swing over to the other side).

- **'Sailing close to the wind'** is sailing nearly into the wind.

- **'Running with the wind'** is sailing with a fast wind behind.

Airships

Rigid envelope shell

Envelope filled with helium

▼ This is a cutaway of one of the new breed of small airships made from lightweight materials and filled with safe, non-flammable helium gas.

Airbags or 'ballonets' inside the helium-filled envelope. As the airship climbs, air pressure drops and the helium expands, pushing air out of the ballonets. As the airship drops again, the helium contracts and air is let into the ballonets again

Valve to let air in and out of the ballonets

Elevator flaps to help climbing or diving

Gondola where pilot sits

Landing wheel

Propeller powered by a motor car engine

Rudder to steer the airship to the left or right

- **By the mid-1800s** ballooning was a popular activity, but balloons have to float where the wind takes them. So in 1852, French engineer Henri Giffard made a cigar-shaped balloon filled with the very light gas hydrogen. He powered it with a steam-driven propeller and added a rudder to make it more 'dirigible' or steerable.

- **In 1884** two French inventors, Charles Renard and Arthur Krebs, built the first really dirigible balloon, *La France*. This was powered by an electric motor.

- **In 1897** Austrian David Schwartz gave a powered, cigar-shaped balloon a rigid frame to create the first airship.

- **In 1900** Count Ferdinand von Zeppelin built the first of his huge airships, the 128 m long LZ-1.

- **In 1909** Zeppelin helped set up the wold's first airline, DELAG, flying 148 m long airships, carrying 10,000 passengers in their first four years.

- **In World War 1** Germany used Zeppelin airships to scout enemy positions and make the first aerial bombing raids.

- **By the 1920s** vast airships were carrying people to and fro across the Atlantic in the style of a luxury ocean-liner. The *Graf Zeppelin* flew at 130 km/h. In its gondola, 60 or more passengers sat in comfortable lounges, walked to cocktail bars or listened to bands playing.

- **On 6 May 1937** disaster struck the giant 245 m-long airship *Hindenburg* as it docked at Lakehurst, New Jersey. The hydrogen in its balloon caught fire and exploded, killing 35 people. The day of the airship was over. Hydrogen was just too dangerous.

- **In recent years** there has been a revival of airships for advertising, filled with safer helium gas. Most are non-rigid, but Airship Industries Skyship is semi-rigid and made from modern light material like carbon-fibre.

★ STAR FACT ★
Fighter planes could take-off from and land on the 1930s airship *Akron* in mid-air.

Underground railways

- **Underground railways** are also called subways, metros or even just tubes.
- **There are three kinds:** open-cut; cut and cover and tube.
- **Open-cut subways** are built by digging rectangular ditches in streets, like much of New York's subway.
- **Cut-and-cover subways** are when an open-cut subway is covered again with a road or pavement.
- **Tubes** are deep, round tunnels created by boring through the ground, like most of London's lines.
- **The world's first underground** was the cut-and-cover Metropolitan Line in London, opened on 10 January 1863, using steam engines.
- **In 1890** the world's first electric tube trains ran on London's first deep tube, the City and South London.
- **After 1900** American Charles Yerkes, builder of the Chicago Loop railway, helped give London the world's most extensive tube network with 225 km of deep tubes.
- **Moscow's Metro** grand stations were built in the 1930s

▲ London was the first city in the world to have an underground system. Much of it is now in deep tube tunnels.

with marble, stained glass, statues and chandeliers.

- **New York City** has the world's largest subway network, but unlike London's most of it is quite shallow. The first line opened on 27 October 1904.

The Eiffel Tower

- **The Eiffel Tower** in Paris was 312.2 m high when it was first built. An antenna brings it up to 318.7 m. There are 1665 steps up to the top.
- **On a clear day** you can see 80 km in all directions from the top. It is often sunny at the top when the weather in the Paris streets is cloudy.
- **It was made** from 18,038 pieces of iron, held together by 2,500,000 rivets.
- **It was built in 1889** for the exhibition celebrating the 100th anniversary of the French Revolution.
- **Gustave Eiffel** (1832-1923) was the most successful engineer of the day, building not only the Eiffel Tower but New York's Statue of Liberty too.

▲ Paris's Eiffel Tower is one of the world's most famous landmarks.

- **The Eiffel Tower** was designed by Maurice Koechlin and Emile Nougier who calculated the effects of wind and gravity with amazing precision.
- **The Tower** was intended to show what could be done with cast iron.
- **During building work** Parisian artists, such as the composer Gounod and poet Maupassant, protested against its ugliness. But when completed it was an instant success with ordinary people.
- **Eiffel** was responsible for financing the construction of the tower, which cost more than $1 million.
- **When Paris was occupied** by the Nazis during the war, the lifts 'mysteriously' stopped working. They restarted the day Paris was liberated.

Missiles

★ STAR FACT ★
American Tomahawk cruise missiles could be aimed through goalposts at both ends of a football field 500 km away.

Explosive warhead

Solid rocket fuel

▶ *90% of the weight of a ballistic missile is the rocket propellant needed to reach its distant target.*

- **In AD1232** the Chinese defended the city of K'ai-feng against the Mongols with gunpowder rockets.
- **In the early 1800s** British army officer William Congreve developed metal rockets carrying explosives.
- **In World War 2** the Germans developed the first guided missiles – missiles steered to their target in flight.

- **The most frightening** German guided missiles were the V-1 flying bombs or 'doodlebugs' and the V-2 supersonic rockets. The V-2 flew at 5300 km/h.
- **Ballistic missiles** arch through the air like a thrown ball. Rockets propel them on the upward trajectory (path). They then coast down on their target. Cruise missiles are propelled by jet on a low flat path all the way.
- **In the 1950s** the USA and Soviet Union competed to develop long-range ICBMs (Intercontinental Ballistic Missiles) usually armed with nuclear warheads.
- **In the 1960s,** antiballistic missiles were developed to shoot down missiles.
- **Some ICBMs** have a range of over 5000 km. Short range missiles (SRBMs) like Pershings reach up to 500 km.
- **SAMS** (surface-to-air missiles) like Redeye are fired from the ground at aircraft. Some can be fired by a soldier with a backpack. AAMs (air-to-air missiles) like Sidewinders are fired from planes against other planes.

Ancient America

- **The largest buildings** in Ancient America were Mayan pyramids, like the Tomb of Pakal in Mexico.
- **Teotihuacan** in Mexico is one of the best preserved ancient cities in the world, with its magnificent pyramids, palaces, temples, courts and homes.
- **At its height** around 2000 years ago, Teotihuacan was the world's biggest city with 100,000 people.
- **Hopewell Indians** of Newark, Ohio built earthwork tombs in AD250 to rival Egypt's pyramids in size.

★ STAR FACT ★
The Incas of Peru built 25,000 km of stone-paved roads between 1450 and 1532.

- **Aztecs** built the Great Temple pyramid in their ancient capital of Tenochtitlan (now Mexico City) from 1325-1500. Human sacrifices were made at the top.
- **The Mesa Verde** is a famous 'pueblos' (stone village) of the Anasazi in New Mexico, abandoned around 1100 AD.
- **In the Chaco Canyon** the Anasazi built 650 km of mysterious roads to nowhere.
- **Nazca Lines** are huge outlines of birds, monkeys and other things up to 100 m across, drawn by Nazca people in the desert in S. Peru. They date from c. 100BC to AD700.
- **The Nazca Lines** are only clearly seen from the air. They may have been part of a giant astronomical calendar.

◀ *The 1500 year-old pyramids of the Maya were once overgrown by jungle, but most have now been restored.*

Palaces

- **The palaces** of Europe owe much to the vast palaces of the caliphs of the Near East, built in 7th and 8th centuries, with their cool courtyards and rich decoration.

- **The Topkapi Palace** in Istanbul was the sumptuous home of the Ottoman sultans for over 400 years until it was made into a museum in 1924, after the sultans' fall.

- **Venice's Doge's Palace** (mainly 1400s) shows oriental touches bought from the east by Venetian merchants.

- **The elegant Pitti Palace** in Florence was designed by the brilliant architect Brunelleschi for the merchant Luca Pitti in 1440. It was taken over by the Medici family in 1550.

- **Hampton Court Palace** southwest of London was begun in 1514 by King Henry VIII's favourite Cardinal Wolsey – but when Wolsey fell from favour, the King took it over.

- **The Escorial** in Madrid is a massive granite palace built 1563-84 by Phillip II to celebrate victories over the French.

- **Many 18th-century princes** built their own Versailles – lavish palaces like Dresden and Potsdam in Germany,

▲ London's Buckingham Palace is the home of the Queen of Britain.

Vienna's Schönbrunn and St Petersburg's Winter Palace.

- **Buckingham House** was built in 1702-5 for the Duke of Buckingham but remodelled as a palace in 1825.

- **Inspired by English gardens** Peter the Great made Russia's oldest garden, the Summer Garden in St Petersburg in 1710. He built the Summer Palace in it.

- **Catherine the Great** had the St Petersburg's Winter Palace built by architect Bartolomeo Rastrelli in 1762.

First steamships

- **In 1783** French nobleman Marquis Claude de Jouffroy d'Abbans built a massive steam boat that churned up the Saone River near Lyon in France for 15 minutes before the pounding engines shook it to bits.

- **In 1787** American John Fitch made the first successful steamship with an engine driving a series of paddles.

- **In 1790** Fitch started the world's first steam service on the Delaware River.

- **In 1802** Scot William Symington built a tug, the *Charlotte Dundas*, able to tow two 70-tonne barges.

- **In 1807** American Robert Fulton made the first steam passenger boats, running 240 km up the Hudson River from New York.

- **In 1819** the New York-built ship *Savannah* made the first steam-assisted crossing of the Atlantic.

- **In 1833** Canadian ship *Royal William* made the first mainly steam-powered Atlantic crossing in two days.

- **In 1843** British engineer Isembard Kingdom Brunel launched the first all-iron hull steamship, the *Great Britain*.

- **In 1858** Brunel launched *Great Eastern*, the biggest ship of the 19th century, 211 m long and weighing 30,000 tonnes.

- **Early steamships** had paddles, but in 1835 Swede John Ericsson invented a screw propeller. In 1845 a screw-driven boat won a tug-of-war with a paddle steamer.

◄ Warships like this, HMS Theseus, were built in the 1880s.

Taking off

- **An aircraft's wings** or 'foils' are lifted by the air flowing above and beneath them as they slice through the air.

- **Because the top** of the wing is curved, air pushed over the wing speeds up and stretches out. The stretching of the air reduces its pressure.

- **Underneath the wing** air slows down and bunches up, so air pressure here rises.

- **The wing gains 'lift'** as the wing is sucked from above and pushed from below.

★ STAR FACT ★
Slots on the wing's leading edge smooth
airflow to increase the safe angle of attack.

- **The amount of lift** depends on the angle of the wing – called the angle of attack – and its shape, and also how fast it is moving through the air.

- **Aircraft** get extra lift for climbing by increasing their speed through the air and by dropping the tail so that the main wings cut through the air at a steeper angle.

- **If the angle of attack** becomes too steep, the air flow breaks up and the wing loses lift. This is called a stall.

- **Planes** take off when air is moving fast enough over the wing to provide enough lift.

- **Airliners** have 'high-lift' slots and flaps on the wings to give extra lift for slow take-off and landing speeds.

◄ *The high-lift flaps are down to give extra lift on a climb.*

The Taj Mahal

- **The Taj Mahal** (said *tarj m'harl*) in Agra in India is perhaps the most beautiful tomb in the world.

- **Mughal Indian** ruler Shah Jehan ordered it to be built in honour of his favourite wife Mumtaz Mahal, who died giving birth to their 14th child.

- **Mumtaz** died in 1629, and the Taj was built over 22 years from 1632 to 1653.

- **The Taj** is set at the north end of a formal Persian garden with water courses and rows of cypress trees.

- **It is made of white** marble and sits on a platform of sandstone.

- **Inside** behind an octagonal screen of alabaster marble tracery lie the jewel-inlaid

cenotaphs (tombs) of Mumtaz and Shah Jehan. The Shah was placed there when he died and his tomb is the only asymmetrical feature in the Taj.

- **20,000 workers** worked in marble and sandstone, silver, gold, carnelian, jasper, moonstone, jade, lapis lazuli and coral to enhance the Taj's beauty.

- **At each corner** of the platform is a slender minaret 40.5 m tall.

- **In the centre** is a dome 21.3 m across and 36.6 m high.

- **The main architect** was Iranian Isa Khan, but the decorations were said to be by Austin of Bordeaux and Veroneo of Venice.

◄ *So perfect are the Taj's proportions that it was said to have been designed by giants and finished by jewellers.*

Jet engines

Front fan to create 'cold' bypass stream

The engine casing is made of carbon-fibre and plastic honeycomb for lightness. Inside is an outer bypass duct for the 'cold-stream' of air from the front fan. An inner duct takes the 'hot stream' through the compressor, combustion chamber and turbine to create the exhaust.

Air intake

Exhaust where a hot jet of air roars out

▶ All but the very fastest warplanes are powered by turbofan jet engines, like this Russian MiG. Turbofans first came into widespread use in the 1970s and are now by far the most common kind of jet engine.

- **A kind of jet engine** was built by the Ancient Greek Hero of Alexander in the first century AD. It was a ball driven round by jets of steam escaping from two nozzles.

- **The first jet engines** were built at the same time in the 1930s by Pabst von Ohain in Germany and Frank Whittle in Britain – though neither knew of the other's work.

> ★ STAR FACT ★
> In a typical turbojet, exhaust gases roar from the engine at over 1600 km/h.

- **Ohain's engine** was put in the Heinkel HE-178 which first flew on 27 August 1939; Whittle's was put in the Gloster E28 of 1941. The first American experimental jet was the Bell XP-59 Aircomet of 1942.

- **Jets** work by pushing a jet of air out the back. This hits the air so fast that the reaction thrusts the plane forward like a deflating balloon.

- **Jet engines** are also called gas turbines because they burn fuel gas to spin the blades of a turbine non-stop.

- **Turbojets** are the original form of jet engine. Air is scooped in at the front and squeezed by spinning 'compressor' blades. Fuel sprayed into the squeezed air in the middle of the engine burns, making the mixture expand dramatically. The expanding air not only pushes round turbines which drive the compressor, but also sends out a high-speed jet of hot air to propel the plane. This high-speed jet is noisy but good for ultra-fast warplanes and the supersonic Concorde.

- **Turboprops** are turbojets that use most of their power to turn a propeller rather than force out a hot air jet.

- **Turbofans** are used by most airliners because they are

quieter and cheaper to run. In these, extra turbines turn a huge fan at the front. Air driven by this fan bypasses the engine core and gives a huge extra boost at low speeds.

- **Ramjets** or 'flying stovepipes' are the simplest type of jet engine, used only on missiles. They dispense with both compressor and turbine blades and simply rely on the speed of the jet through the air to ram air in through the intake into the engine.

▶ Like nearly all warplanes today, 'stealth' aircraft are jet-propelled. But the afterburner stream of hot gases from the jets provides a 'signature' that can show up all too clearly on some detection equipment. So stealth aircraft are designed to 'supercruise' – that is, fly at supersonic speeds without much afterburn.

Trains of the future

▲ *Many cities now have short monorails, but they seem unlikely to get much bigger because of the disruption building would cause.*

- **Monorails** are single beam tracks raised over city streets.

- **The first monorail** was built in Wuppertal in Germany as long ago as 1901 and monorails have been seen as trains of the future ever since.

- **Monorails** of the future may be air-cushion trains or maglev, as at Birmingham airport.

- **A maglev** is proposed in Japan to take passengers the 515 km from Tokyo to Osaka in under 60 minutes. Germany is planning a system called Transrapid.

- **PRT** or Personalized Rapid Transport is automatic cars on monorails. AGT (Automated Guideway Transit) is the same for coaches, as in Detroit, or London's DLR.

- **300 km/h plus** High Speed Train (HST) systems like the French TGV are being built in many places, such as from Moscow to St Petersburg.

- **In 2004** the 320 km/h Tampa-Miami Florida Overland Express opens. The same year a 4500 km 300 km/h line may open from Melbourne to Darwin, Australia.

- **Most HSTs** run on special straight tracks. Tilting trains lean into bends to give high speeds on winding old tracks.

- **Tilting trains** include the 300 km/h Italian Fiat Pendolini and the Swedish X2000.

- **From 2000** the 240 km/h tilting train *The American Flyer* – Washington to Baltimore – is the USA's fastest train.

Rockets

· STAR FACT ·
The *Saturn V* rocket that launched the Apollo mission to the Moon is the most powerful rocket ever built.

- **Rockets** work by burning fuel. As fuel burns and swells out behind, the swelling pushes the rocket forward.

- **Solid-fuel rockets** are the oldest of all engines, used by the Chinese a thousand years ago.

- **Solid-fuel engines** are basically rods of solid, rubbery

◄ *Only powerful rockets can give the thrust to overcome gravity and launch spacecraft into space. They fall away in stages once the spacecraft is travelling fast enough.*

fuel with a tube down the middle. When the fuel is set alight it burns out through the fuel until all is used.

- **Solid fuel** rockets are usually only used for model rockets and small booster rockets. But the Space Shuttle has two solid rocket boosters (SRBs) as well as three main liquid fuel engines.

- **Most powerful launch rockets** use liquid fuel. The Shuttle uses hydrogen. Other fuels include kerosene.

- **Liquid fuel** only burns with oxygen, so rockets must also carry an oxidizer (a substance that gives oxygen) such as liquid oxygen (LOX) or nitrogen tetroxide.

- **Future rocket drives** include nuclear thermal engines that use a nuclear reactor to heat the gas blasted out.

- **NASA's Deep Space-1** project is based on xenon ion engines which thrust, not hot gases out the back, but electrically charged particles called ions.

- **Solar thermal engines** of the future would collect the Sun's rays with a large mirror to heat gases.

Navigation

- **Early sailors** found their way by staying near land, looking for 'landmarks' on shore. Away from land they steered by stars, so had only a vague idea of direction in the day.

- **After 1100** European sailors used a magnetic compass needle to find North.

- **A compass** only gives you a direction to steer; it does not tell you where you are.

- **The astrolabe** was used from c.1350. This measured the angle of stars above the horizon, or the Sun at noon giving an idea of latitude (how far north or south of the equator).

- **From the 1500s** the cross-staff gave a more accurate measure of latitude at night from the angle between the Pole Star and the horizon.

- **From the mid-1700s** until the 1950s, sailors measured latitude with a mirror sextant. This had two mirrors. It gave the angle of a star (or the Sun) when

◀ A navigational instrument for measuring latitude from the angle of certain stars.

one mirror was adjusted until the star was at horizon height in the other.

- **For centuries** the way to find longitude – how far east or west – was by dead reckoning. This meant trailing a knotted rope in the water to calculate speed, and so estimate how far you had come.

- **You can find longitude** by comparing the Sun's height with its height at the same time at a longitude you know. But early pendulum clocks did not work well enough aboard ship to give the correct time.

- **The longitude problem** was solved when John Harrison made a very accurate spring-driven clock or chronometer.

- **Ships** can now find their position with pinpoint accuracy using the Global Positioning System or GPS. This works by electronically comparing signals from a ring of satellites.

Sydney Opera House

▲ Standing by the water's edge in Sydney Harbour, Sydney Opera House is one of the world's most distinctive buildings.

- **Sydney Opera House** in Sydney, opened in October 1973, is one of Australia's best known landmarks.

- **The design** by Danish architect Jorn Utzon was agreed in January after a competition involving 233 entries, but it took 14 years to build.

- **The radical sail-like** roofs were said to have been inspired by sea shells.

- **The original design** was so unusual that it proved impossible to build structurally, so Utzon altered the shape of the roofs to base them on sections of a ball.

- **The altered design** meant the roofs could be made from pre-cast concrete slabs.

- **The roofs** were made from 2194 pre-cast concrete slabs, weighing up to 15 tonnes each.

- **The slabs** are tied together with 350 km of tensioned steel cable and covered with 1 million Swedish tiles.

- **In the mouths of the roofs** are 6225 sq m of French glass in double layers, one tinted topaz-colour.

- **Inside** there were originally four theatres – the Concert Hall, the Opera Theatre, the Drama Theatre and the Playhouse.

- **A new theatre, The Studio,** was added in March 1999 for small cast plays and films.

Bombers of World War II

- **In the 1930s** Boeing built the B-17 'Flying Fortress', with gun turrets to battle its way through to targets even by day.

- **The 1929** Curtis F8C Helldiver was the first 'dive-bomber' – designed to drop its bombs at the end of a long dive on targets like aircraft carriers. German 'Stuka' dive-bombers gained a fearsome name in the German invasions of 1939.

- **The twin-engined** Heinkel 111, Dornier Do17 and Junkers Ju88 were the main German bombers in the *Blitzkrieg* (literally 'lightning war') raids of the Battle of Britain.

- **In December 1939** the heavy loss of British Wellingtons showed that lightly armed bombers could not sustain daylight raids, so the British switched to night raids.

- **Blind bombing** radar systems like the Hs2, and flare trails left by advance 'Pathfinder' missions, improved accuracy on night raids.

◀ The British Avro Lancaster could carry 6000 kg of bombs on low altitude raids.

- **The ultra-light** De Havilland Mosquito was fast enough to fly daylight raids.

- **The Russian Ilyushin Il-2** or Stormovik was so good at bombing tanks Stalin said it was 'as necessary to the Red Army as air or bread.'

- **The dambusters** were the Lancasters of 617 squadron of 1943 that attacked German dams with 'bouncing bombs'. These were round bombs designed by Barnes Wallis that bounced over the water surface towards the target dams.

- **Kamikaze** (Japanese for 'divine wind') were fighters loaded with bombs and gasoline which their pilots aimed in suicide dives at enemy ships.

- **The biggest bomber** was the Boeing B-29 Superfortress which could fly over 10,000 m up. In 1945, B-29s dropped the atomic bombs on Hiroshima and Nagasaki in Japan.

Early cars

- **In 1890** Frenchman Emile Levassor made the first real car, with an engine at the front. He laughed, saying, *C'est brutal, mais ca marche* ('It's rough, but it goes'.)

- **The Duryea** brothers made the first successful American car in 1893.

- **Early accidents** made cars seem dangerous. Until 1896 in Britain and 1901 in New York cars had to be preceded by a man on foot waving a red flag.

- **In 1895** the French Panhard-Levassor company made the first covered 'saloon' car.

- **In 1898** Renault drove the wheels with a shaft not a chain.

- **The Oldsmobile** Curved Dash of 1900 was the first car to sell in thousands.

- **On early cars** speed was controlled by moving a small ignition advance lever backwards or forwards.

- **Dirt roads,** oil spray and noise meant early motorists needed protective clothing such as goggles and earmuffs.

- **The first cars** had wooden or wire spoked wheels. Pressed steel wheels like today's came in after 1945.

- **In 1906** an American steam-driven car, the Stanley Steamer, broke the land speed record at over 205 kmh.

◀ By 1904, many cars were starting to have the familiar layout of cars today – engine at the front, driver on one side, steering wheel, petrol tank at the back, a shaft to drive the wheels and so on.

Canals

- **In 1470** BC the Egyptian Pharaoh Sesostris had the first Suez Canal built, linking the Mediterranean and Red Sea.

- **The Grand Canal** in China is the world's longest man-made waterway, running 1747 km west of Beijing.

- **The origins** of the Grand Canal date to the 4th century BC, but it was rebuilt in 607 AD.

- **The late 1700s and early 1800s** saw many canals built for the factories of the Industrial Revolution, like James Brindley's Bridgwater Canal in Lancashire, England.

- **Clinton's Folly** is the Erie Canal linking Lake Erie to the Hudson in the USA, pushed through by governor De Witt Clinton in 1825.

- **The current Suez Canal** was built by Ferdinand de Lesseps (1859-1869). It is the world's longest big-ship canal, 161.9 km long.

- **The Panama Canal** links the Atlantic and Pacific Oceans 82 km across Central America, and cutting the sea journey from New York to San Francisco by 14,400 km.

▲ Amsterdam's canals were dug in the city's heyday in the 1600s.

- **The Panama Canal** takes over 400 million tonnes of shipping a year – more than any other canal.

- **The world's busiest canal** is Germany's Kiel which takes 45,000 ships a year from the North Sea to the Baltic.

- **In Russia** one of the world's longest canal systems links the Black Sea to the Arctic Ocean via the Volga River.

Airliners

- **The Boeing 247** of 1933 was the world's first modern airliner, with smooth monoplane wings, streamlined metal body and retractable landing wheels.

- **The Douglas DC-3** of 1936 could carry 21 passengers smoothly at 320 km/h and was the first popular airliner.

- **In 1952** the world's first jet airliner, the De Havilland Comet 1 came into service.

- **The Comet** more than halved international flight times but several tragic accidents led to its grounding in 1954.

▶ The four-engined Boeing 747 flies at 10-13,000 m – well above most storms – and can fly non-stop from New York to Tokyo.

> **! NEWS FLASH !**
> Spaceplanes like Lockheed-Martin's *Venture Star* may make space trips routine flights.

- **The age of jet** air travel really began with the American Boeing 707 and Douglas DC-8 of the late 1950s.

- **The Boeing 747** jumbo jet of 1970 – the first 'wide-bodied jet' – had over 400 seats, making air travel cheap.

- **Four-engined jets** like the 747 can fly 10,000 km non-stop at speeds of 1000 km/h. Two and three-engine jets like the DC-10 make shorter flights.

- **Supersonic airliners** able to travel at over 2000 km/h like the Anglo-French Concorde and Russian Tupolev Tu-144 have proved too heavy on fuel and too noisy.

- **The planned 555-seat Airbus** A380 would be the first full-length double-deck airliner.

GEOGRAPHY

KEY

 Asia

 The Americas

 Europe

 Africa and Australasia

 People

 Places

The Alps

▲ *The pointed summit of the Matterhorn is the third highest peak of the Alps.*

- **The Alps** are Europe's largest mountain range, 1050 km long, up to 250 km wide and covering 210,000 sq km.
- **The highest Alpine peak** is Mont Blanc (4807 m) on the France-Italy border.
- **Famous peaks** include the Matterhorn (4478 m) and Monte Rosa (4634 m) on the Swiss-Italian border.

- **The Alps began to form** about 65 million years ago when the African crustal plate shifted into Europe.
- **The Alps are the source** of many of Europe's major rivers such as the Rhone, Po and Rhine.
- **Warm, dry winds** called föhns blow down leeward slopes, melting snow and starting avalanches.
- **The high Alpine pastures** are famous for their summer grazing for dairy cows. In winter, the cows come down into the valleys. This is called transhumance.
- **The Alps** are being worn away by human activity. In valleys, cities and factories are growing, while skiing wears away the slopes at the tops of the mountains.
- **The Alps** have Europe's highest vineyards, 1500 m up.

Thailand and Myanmar

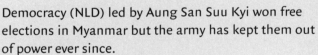

▶ *Many canals thread through Bangkok and provide a way for poor people to bring goods to sell.*

- **Thailand:** Capital: Bangkok. Population: 60.8 million. Currency: Baht. Language: Thai.
- **Myanmar:** Capital: Rangoon. Population: 49.4 million. Currency: Kyat. Language: Burmese.
- **In 1990** the National League for Democracy (NLD) led by Aung San Suu Kyi won free elections in Myanmar but the army has kept them out of power ever since.
- **Growing opium poppies** to make the painkiller morphine and the drug heroin is one of the few ways the people of north Myanmar can make money.
- **The world's best rubies** come from Mogok in Myanmar.
- **Thailand's capital** is called Bangkok by foreigners but its real name has over 60 syllables. Many locals call it Krung Thep which are the first two syllables.
- **Most people in Thailand** and Myanmar still live in the countryside growing rice to eat.
- **Most people** live on fertile plains and deltas – around the Irrawaddy River in Myanmar and the Chao Phraya in Thailand.
- **Millions** of tourists come to Thailand each year to visit the country's beaches, and the attractions of the city of Bangkok.

Greece

- **Capital:** Athens. Area: 131,957 sq km. Currency: euro. Language: Greek.
- **Physical features:** Highest mountain: Mt Olympus (2,917 m).
- **Population:** 10.6 million. Population density: 80/sq km. Life expectancy: men 76.0 years; women 81.3 years.
- **Wealth:** GDP: $130.6 billion. GDP per head: $12,320.
- **Exports:** Clothes, olive oil, petroleum products, fruit and tobacco.
- **Farming:** Greece is so mountainous that less than a third can be cultivated, but a fifth of all workers work on the land, many raising sheep or growing olives or vines for wine.

▼ Greece is one of the most mountainous countries in Europe.

- **Greece** is the world's third largest grower of olives after Spain and Italy.
- **Greek salad** includes olives and feta cheese from goats milk. In some small villages, bakers allow villagers to cook their food in their *fuorno* oven.
- **More than 11 million** visitors come to Greece each year – some to see the relics of Ancient Greece, but most to soak up the sun.
- **Athens** is the ancient capital of Greece, dominated by the Acropolis with its famous Parthenon temple ruins. Athens is also a modern city, with pollution caused by heavy traffic.

World trade

- **International trade** is the buying and selling of goods and services between different countries.
- **International trade** has increased so much people talk of the 'globalization' of the world economy. This means that goods are sold around the world.
- **The balance of world trade** is tipped in favour of the world's richest countries and companies.
- **Just 200 huge multinational** companies control much of world trade.
- **Just five countries** – the USA, Germany, Japan, France and the UK – control almost half world trade.
- **The 30 richest countries** control 82% of world trade.
- **The 49 poorest countries** control just 2% of world trade.
- **Some countries** rely mainly on just one export. 95% of Nigeria's export earnings come from oil; 75% of Botswana's come from diamonds.

- **Some countries** such as the USA want 'free trade' – that is, no restrictions on trade; other less powerful nations want tariffs (taxes on foreign goods) and quotas (agreed quantities) to protect their home industries.
- **The World Trade Organization** was founded on Jan 1, 1995 to police world trade, and to push for free trade.

▼ This diagram shows the proportions of each kind of good traded around the world.

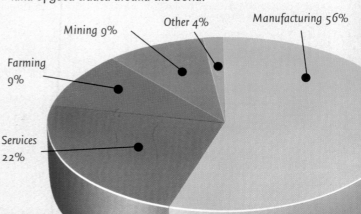

Mining 9%
Other 4%
Manufacturing 56%
Farming 9%
Services 22%

Peoples of northern Asia

- **86% of Russians** are descended from a group of people called Slavs who first lived in eastern Europe 5,000 years ago.

- **East Slavs** are the Great Russians (or Russians), the Ukrainians and the Belarusians (or White Russians).

- **West Slavs** are eastern Europeans such as Czechs, Poles and Slovaks.

- **South Slavs** are

▲ The Kazakhs or Cossacks were famous for their horseriding skills.

people such as Croats, Serbs and Slovenes.

- **Slavs speak** Slavic or Slavonic languages such as Russian, Polish or Czech.

- **In the old Soviet Union** there were over 100 ethnic groups. 70% were Slavs. Many of the rest were Turkic people such as Uzbeks, Kazakhs and Turkmen. Many of these peoples now have their own nations.

- **Slavic** people are mainly Christian; Turkic people are mainly Islamic.

- **Many Turkic peoples** such as the Kazakhs have a nomadic tradition that is fast vanishing.

- **The Mongols** were a people whose empire under the great Khans once spread far south into China and far west across Asia.

- **The Tatars** are 4.6 million Turkic people who now live mainly in the Tatar Republic in the Russian Federation.

Romania and Bulgaria

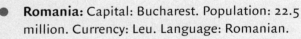 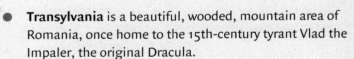

- **Romania:** Capital: Bucharest. Population: 22.5 million. Currency: Leu. Language: Romanian.

- **Bulgaria:** Capital: Sofia. Population: 8.3 million. Currency: Lev. Language: Bulgarian.

- **Romania** gets its name from the Romans who occupied it almost 2,000 years ago.

- **Transylvania** is a beautiful, wooded, mountain area of Romania, once home to the 15th-century tyrant Vlad the Impaler, the original Dracula.

- **Like Bulgaria,** Romania was communist until 1989 when the people overthrew President Ceausescu.

- **Ceausescu's** attempts to develop industry forced people off the lands into towns. Many orphans were left as families broke up.

- **Romania** is a major wine grower.

- **Romania** is home not only to native Romanians but 400,000 Roma (gypsies).

- **The Valley of Roses** is a valley near Kazanluk in Bulgaria full of fields of damask roses.

- **Bulgarian women pick** damask rose blossoms to get the oil to make 'attar of roses', used for perfumes.

 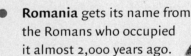

◄ Damask rose blossoms in Bulgaria's Valley of Roses are picked early mornings in May and June, before the sun dries out the petals.

DR Congo

- **Capital:** Kinshasa. Area: 2,344,856 sq km. Currency: Congolese franc. Official language: French.
- **Physical features:** Highest mountain: Mont Ngaliema (5109 m). Longest river: Congo (4667 km).
- **Population:** 51.8 million. Population density: 22/sq km. Life expectancy: men 49 years; women 52 years.
- **Wealth:** GDP: $5 billion. GDP per head: $110.
- **Exports:** Copper, diamonds, coffee, cobalt, petroleum.
- **DR Congo** is called Democratic Republic of Congo to distinguish it from a neighbouring country

▲ Congo lies on the Equator. Over a third of it is thick equatorial rainforest.

also called Congo, Congo (Brazzaville).

- **Once called the** Belgian Congo DR Congo then became the Congolese Republic, then Zaire. Since 1997 it has been the Democratic Republic of Congo
- **Congo** is one of the world's leading copper producers. There is a vast copper mine in Shaba.
- **Congo is** the world's leading industrial diamonds producer.
- **The Congo River** is the world's ninth longest river, and carries more water than any river but the Amazon.

Southern USA

- **The south central states** such as Texas, Oklahoma and New Mexico produce a lot of oil and gas.
- **Texas** produces more oil than any other state apart from Alaska.
- **One of the world's** largest oil companies, was founded on Texan oil.
- **Texas** is known as the Lone Star state.

> ★ STAR FACT ★
> The first integrated circuit was invented in 1958 in Dallas

- **Louisiana** is known as the Sugar state because it grows so much sugar.
- **Oil wealth and aerospace** have attracted high-tech industries to Texan cities like Dallas, Houston and San Antonio.
- **Cotton** is grown on the Mississippi plains, while tobacco is important in the Carolinas and Virginia.
- **In the mid 1800s** the southern states grew 80% of the world's cotton, largely using black slave labour.
- **Florida** is famous for Disneyworld, the Cape Canaveral space centre and the Everglades, a vast area of steamy tropical swamp infested by alligators.

◀ Texas's wealth came with the discovery of oil in 1901. Now aerospace and high-tech industries are thriving in this sunny state.

Zimbabwe

▶ *Victoria Falls on the Zambezi is one of the world's biggest waterfalls. Its roar can be heard 40 km away.*

- **Capital:** Harare. Area: 390,759 sq km. Currency: Zimbabwe dollar. Official language: English.

- **Physical features:** Highest mountain: Mt Inyangani (2592 m). Longest river: Zambezi (3540 km).

- **Population:** 12.4 million. Population density: 30/sq km. Life expectancy: men 39 years; women 40 years.

- **Wealth:** GDP: $6.3 billion. GDP per head: $520.

- **Exports:** Tobacco, ferrochrome, textiles and clothing, nickel.

- **Zimbabwe** was once the British colony of Rhodesia but was granted independence in 1980.

- **The name Zimbabwe** came from the huge ancient stone palace of Great Zimbabwe (which means 'house of stone').

- **Zimbabwe** is a fertile farming country, growing lots of tobacco, cotton and other crops. Much of the land still remains in the hands of white farmers, but the government plans to change this situation.

- **Zimbabwe** is the most industrial African nation after S. Africa, making steel, cement, cars, machines, textiles and much more. Harare is the biggest industrial centre.

- **98% of Zimbabweans** are black. The Shona people are the biggest group, then come the Ndebele (or Matabele). The Shona speak a language called Shona, the Ndebele speak Matabele.

Chinese food

- **The staple foods** in China are rice and wheat with corn, millet and sorghum. In the south, the people eat more rice. In the north, they eat more wheat, as bread or noodles.

- **Vegetables** such as cabbage, bean and bamboo shoots are popular. So too is *tofu* (soya bean curd).

- **Favourite meats** in China are pork and poultry, but the Chinese also eat a lot of eggs, fish and shellfish.

- **A Chinese breakfast** may be rice and vegetables or rice porridge and chicken noodle soup or sweet pastries.

- **A Chinese lunch** may include egg rolls or meat or prawn dumplings called *dim sum*.

◀ *A favourite snack in China is fried savoury dumplings.*

- **A Chinese main meal** may be stir-fried vegetables with bits of meat or seafood in a stock, with rice or noodles.

- **China has** a long tradition of fine cooking, but styles vary. Cantonese cooking in the south has lots of fish, crabs and prawn. Huaiyang has steamed dishes. Sichuann is spicy. Beijing cooking in the north is the most sophisticated, famous for its Peking duck (cripsy roast duck).

- **The Chinese** often cook their food by stir-frying (stirring while hot frying) in big round pans called woks. They eat the food from bowls with chopsticks and small china spoons, not with knives and forks.

- **Chinese** drink tea without milk, typically made from jasmine leaves, oolong (green tea) or chrysanthemum.

★ STAR FACT ★
The Chinese were drinking tea at least 4,000 years ago.

Russia

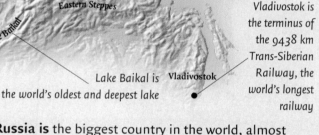

Murmansk
St Petersburg
Barents Sea
Arctic Ocean
Kara Sea
MOSCOW
RUSSIA
Western Steppes
Ural Mountains
Black Sea
Caucasus Mts
Mt Elbrus
Caspian Sea
Volgograd
SIBERIA
Omsk
Sayan Mountains
Irkutsk
Lake Baikal
Eastern Steppes
Sea of Okhotsk
KAMCHATKA
Vladivostok

In eastern Siberia, milk is sold in frozen blocks with a wooden handle

Vladivostok is the terminus of the 9438 km Trans-Siberian Railway, the world's longest railway

Lake Baikal is the world's oldest and deepest lake

▶ Russia stretches 10,000 km – almost a third of the way around the world – from the open steppes south of Moscow to the chilly pine forests of the Kamchatka in the east.

- **Capital:** Moscow. Area: 17,075,400 sq km. Currency: Rouble. Language: Russian.

- **Physical features:** Highest mountain: Mt Elbrus (5642m). Longest river: Yenisey-Angara (5540 km).

- **Population:** 147.7 million. Population density: 9/sq km. Life expectancy: men 60.7 years; women 72.9 years.

- **Wealth:** GDP: $330 billion. GDP per head: $2270. Exports: fuels and lubricants, metals, machinery, transport equipment.

- **Russia** or the Russian Federation is the country created by the Russians after the break up of the Soviet Union in 1991. It includes republics such as Chechnya, Osetiya, Kalmykiya, Tatarstan, Mordoviya and Bashkortostan. Many of these republics, such as Chechnya, are waiting to be independent.

▼ The Ural mountains run from north to south forming a natural boundary between Russia and Siberia, Europe and Asia.

- **Russia is** the biggest country in the world, almost twice as big as the next country, Canada. It stretches from the subtropical south to the Arctic north, where it has the longest Arctic coastline of any country.

- **Russia has huge** mineral resources and is among the world's leading producers of oil, natural gas, coal, asbestos, manganese, silver, tin and zinc. It also has giant forests for timber in the east in Siberia.

- **Russia** has some of the biggest factories in the world around Moscow, producing everything from high-tech goods to iron and steel and trucks. Yet in the far north and east, people still live as they have done for many thousands of years, herding reindeer or hunting.

- **After the USSR broke up,** Russia and its people were plunged into crisis. Encouraged by western nations, Russian presidents – first Boris Yeltsin, then Vladimir Putin – have tried to establish a free market economy in place of the old communist one.

★ **STAR FACT** ★
In Yakutsk in Siberia, winter temperatures can plunge to -69°C while summers can soar to 39°C – more extreme than anywhere else. Oymyakon is the world's coldest village. It once had temperatures of -72°C.

New Zealand

◄ *New Zealand is made up of two main islands – the almost sub-tropical North island where most people live and the long, narrow South island with its wide Canterbury Plains and soaring Southern Alps.*

- **Capital:** Wellington. Area: 270,534 sq km. Currency: NZ dollar. Language: English.

- **Physical features:** Highest mountain: Mt Cook (3754 m). Longest river: Waikato (425 km).

- **Population:** 3.8 million. Population density: 14/sq km. Life expectancy: men 74.3 years; women 79.9 years.

- **Wealth:** GDP: $74.7 billion. GDP per head: $19,660.

- **Exports:** Meat, milk, butter, cheese, wool, fish, fruit.

- **New Zealand** was one of the last places to be inhabited by humans and remains a clean, beautiful land, with rolling farmland, thick forests and towering mountains.

- **New Zealand** is mainly a farming country, with 64% of the land devoted to crops and pasture for sheep and cattle. 60% of New Zealand's exports are farm produce.

- **Fast-flowing** rivers provide 61% of New Zealand's power through hydroelectric plants. Geothermal energy from hot springs provides some of the rest. Nuclear power is banned.

- **The first** inhabitants of New Zealand were the Maoris, who came about AD900 and now form 14.2% of the population. The remaining 85.8% are mostly descended from British and Irish settlers who came in the 19th and 20th centuries.

> ★ **STAR FACT** ★
> There are 45 million sheep in New Zealand – 12 sheep to every person!

Yellowstone Park

- **Yellowstone** is the oldest and best-known national park in the USA. It was established by Act of Congress on March 1, 1872.

- **It is one of the world's largest** parks covering 8987 sq km of rugged mountains and spectacular deep valleys.

- **It is situated** across Wyoming, Montana and Idaho.

- **Yellowstone** is famous for its lakes and rivers such Yellowstone Lake and Snake River.

- **Most of Yellowstone** is forested in lodgepole pines, along with other conifers, cottonwoods and aspens. It also has a wealth of wild flowers.

▼ *Yellowstone sits on top of a volcanic hot spot which gives it its famous geysers and hot springs – and may make it the site of the biggest eruption of all time.*

- **Yellowstone's** wild animals include bison, elk, bighorn sheep, moose, grizzly bears and wolves.

- **Yellowstone** has the world's greatest concentration of geothermal features including 10,000 hot springs and 300 geysers, as well as steam vents, mud cauldrons, fumaroles and paint pots.

- **The most famous geyser** is Old Faithful, which spouts every hour or so. The biggest is the 115 m Steamboat.

- **One of the biggest** volcanic eruptions ever occured in Yellowstone Park two million years ago. Enough lava poured out in one go to build six Mt Fujiyamas.

- **There are signs** that Yellowstone may soon erupt as a 'supervolcano' – an eruption on an unimaginable scale.

International organizations

◄ The Red Cross was set up by Swiss Jean Dunant in the 19th century after he witnessed the bloody slaughter at the battle of Solferino in Italy. It now plays a vital role in helping suffering people everywhere.

- **International organizations** are of three main types: those set up by governments, like the UN; multinationals; and human rights and welfare organizations like the Red Cross and Amnesty International.

- **The United Nations** or UN was formed after World War II to maintain world peace and security. It now has over 190 member nations.

- **UN headquarters** are in New York City. The name was coined by US President Roosevelt in 1941.

- **All UN members** meet in the General Assembly. It has five permanent members (Russia, USA, China, France and UK) and ten chosen every two years.

- **The UN** has agencies responsible for certain areas such as children (UNICEF), food and farming (FAO), health (WHO), science (UNESCO) and nuclear energy (IAEA).

- **Multinationals** or TNCs (transnational corporations) are huge companies that work in many countries.

- **TNCs** like Coca-Cola and Kodak are well known; others like cigarette-makers Philip Morris are less known.

- **Some TNCs** take in more money than most countries. Just 500 TNCs control 70% of all the world's trade.

- **90% of world** grain is handled by six big US TNCs. Cargill and Continental alone control half the world's grain.

- **Amnesty International** was founded in 1961 to campaign for those imprisoned for religious and political beliefs.

Italy

- **Capital:** Rome. Area 301,277 sq km. Currency: euro. Language: Italian.

- **Physical features:** Highest mountain: Monte Bianco di Courmayeur (4760 m). Longest river: Po (652 km).

- **Population:** 57.3 million. Population density: 190/sq km. Life expectancy: men 75.4 years; women 82.1.

- **Wealth:** GDP: $1240 billion. GDP per head: $21,650.

- **Exports:** Wine, machinery, cars and trucks, footwear, clothes, olive oil, textiles, mineral products.

- **Italy** is a narrow, mountainous country. The north is cool and moist, with big industrial cities. Tuscany and Umbria have rich farmland and ancient cities famous for their art treasures. The south is hot, dusty and often poor.

- **Vines and olives** are grown widely and Italy is one of the world's main producers of both wine and olive oil.

- **Italy is one of the biggest** industrial nations. Industry is concentrated in the north in cities like Turin and Milan.

▲ The ancient city of Venice is set on 117 islands in a lagoon. Instead of streets, there are 177 canals, plied by boats called gondolas.

- **Italians** like to dress in style and the fashion trade is big business. Italian fashion labels like Armani, Versace, Valentino, Moschino and Gucci are now world famous.

- **Italy is full of beautiful** historic towns like Florence, Padua and Mantua, many dating from the Renaissance.

Rich and poor

▶▼ *Expensive cars and fine foods are often seen as status symbols for the wealthy.*

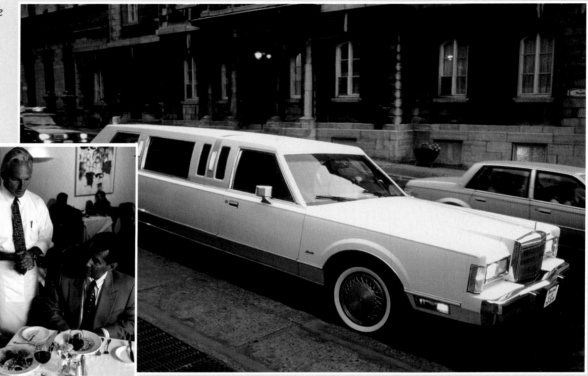

- **The world's richest country** is the USA, with a GDP of $8650 billion ($31,330 per head). But Luxembourg has an even higher GDP per head – $45,320.

- **The world's poorest country** by GDP per head is Sierra Leone. Each person has, on average, $130, but many are even poorer.

- **The world's richest countries** with less than a quarter of the world's population take three-quarters of its wealth.

- **Most of the world's rich countries** are in the Northern Hemisphere. Most poor countries are in the South. So people talk of the North-South divide.

- **One billion people** around the world live in 'absolute poverty'. This means they have no real homes. In cities, they sleep rough or live in shacks. They rarely have enough to eat or drink.

- **In the 1970s** richer countries encouraged poorer countries like Mexico and Brazil to borrow money to build new dams and industrial works.

- **By 1999** poor countries were paying $12 in debt interest for every $1 rich countries were donating in aid.

- **Famine** has become a common problem in the poorer parts of the world. One reason is that so much farmland is used for growing crops for export – raising the cost of food, and restricting the land available for growing food for local people.

- **250,000** children die a week from a poor diet. 250,000 die a month from diarrhoea, because of a lack of clean water.

◀ *Since 1960, the divide between North and South has grown wider leaving many in abject poverty.*

Israel

- **Capital:** Jerusalem. Area: 20,770 sq km. Currency: New Shekel. Languages: Hebrew, Arabic.

- **Physical features:** Highest mountain: Mt Meron (1208 m). Longest river: River Jordan (325 km).

- **Population:** 6.1 million. Population density: 299/sq km. Life expectancy: men 75 years; women 79.9 years.

- **Wealth:** GDP: $107.2 billion. GDP per head: $17,570.

- **Exports:** Diamonds, chemicals and chemical products, fruit and vegetables, machinery.

- **Israel** was founded in 1948 as a home for Jews who have since come here from all over the world.

- **The city of Jericho** may be the oldest in the world, dating back more than 10,000 years.

- **Some people** in rural areas work on kibbutzim – collective farms where work and profits are shared.

- **Israel is famous** for its Jaffa oranges, named after Jaffa, the old name for a town close to Tel Aviv. They are grown on the Plain of Sharon.

> ★ **STAR FACT** ★
> Jerusalem is sacred for three major religions: Judaism, Islam and Christianity.

Rome

- **Rome** is the capital of Italy, and its biggest city, with a population of almost three million.

- **Rome's Vatican** is the home of the Pope.

- **The Vatican** is the smallest independent country in the world covering just 0.4 sq km.

- **Rome is known** as the Eternal City because of its importance within the Roman Empire.

- **Ancient Rome ruled** much of Europe and the lands around the Mediterranean for hundreds of years as the capital of the Roman Empire.

- **Ancient Rome** was famously built on seven hills – the Aventine, Caelian, Capitoline, Esquiline, Palatine, Quirinal and Viminal.

- **Rome has** one of the richest collections of art treasures and historic buildings in the world. The Trevi is one of many beautiful fountains.

- **There are many Ancient Roman** relics in Rome including the Colosseum arena and the Pantheon.

▲ St Peter's Church is located in the Vatican city in Rome.

- **The Vatican's** Sistine Chapel has a ceiling painted brilliantly by Michelangelo and frescoes (wall paintings) by Botticelli, Ghirlandaio and Perugino.

- **Rome is** now a major centre for film-making, publishing and tourism.

Peoples of Africa

- **Africa has been** inhabited longer than any other continent. The earliest human fossils were found here.

- **In the north** in countries such as Algeria, Morocco and Egypt, people are mainly Arabic.

- **The Berber people** were the first people to live in northwest Africa, with a culture dating back to at least 2400BC. Their culture survives in remote villages in the Atlas mountains of Algeria and Morocco.

- **Tuaregs** are camel-herding nomads who live in the Sahara desert, but much of their traditional grazing land has been taken over by permanent farms.

- **South of the Sahara** most people are black Africans.

- ◀ *Zulus are Bantu-speaking people who live in South Africa. They have a proud warrior tradition.*

- **There are more than 3000** ethnic groups of black Africans.

- **Over 1,000** different languages are spoken in Africa.

- **Most people** in southern Africa speak English or one of 100 Bantu languages such as Zulu or Swahili.

- **Many people** in rural southern Africa live in round houses.

- **Africa was ruled** by the Europeans as colonies. By the early 20th century the country was divided into nations. Many small groups became dominated by tribes and cultures perhaps hostile to their own.

Turkey and Cyprus

- **Turkey:** Capital: Ankara. Population: 65.7 million. Currency: Turkish Lira. Language: Turkish.

- **Cyprus:** Capital: Nicosia. Population: 757,000. Currency: Cyprus pound. Languages: Greek and Turkish.

- **Turkey lies** partly in Europe, partly in Asia. The two continents are separated by a narrow sea called the Bosphorus.

- ◀ *A hubble-bubble is a special pipe popular in Turkey. Sucking on the long pipe draws the smoke bubbling through water.*

- **Turkey is a republic** with a mix of Islamic and Western traditions.

- **Istanbul** is one of the world's great historic capital cities. As Byzantium, it was capital of the Byzantine Empire for 1,000 years. Then it was Constantinople, the capital of the great Ottoman Empire for 500 years, until 1923.

- **Street cafés** are popular with Turkish men, who come to drink thick, dark, sweet Turkish coffee, smoke pipes called hubble-bubbles and play backgammon.

- **An estimated 25 million Kurds** live on the borders of Turkey, Iran, Iraq and Syria and have no country of their own.

- **31% Turkish people** live in the country growing wheat, cotton, tobacco, sugar beet, fruit and tea.

- **Turkey's national motto** is *Yurtta sulh, Cihand sulh* ('Peace at home, peace in the world').

- **Turkish food** is famous for its *shish kebabs* – cubes of meat and vegetables barbecued on a skewer.

Venezuela & neighbours

▲ Venezuela and its neighbours lie along the northern coast of South America, along the edge of the tropical waters of the Caribbean.

- **Venezuela:** Capital: Caracas. Population: 24.2 million. Currency: Bolivar. Language: Spanish.

- **Colombia:** Capital: Bogota. Population: 42.3 million. Currency: Colombian peso. Language: Spanish.

- **Guyana:** Capital: Georgetown. Population: 875,000. Currency: Guyana dollar. Official language: English.

- **Suriname:** Capital: Paramaribo. Population: 452,000. Currency: Suriname guilder. Language: Dutch.

- **French Guiana:** Capital: Cayenne. Population: 173,000. Currency: euro. Language: French.

- **The discovery of oil** in Venezuela's Lake Maracaibo in 1917 turned it from one of South America's poorest countries to one of its richest.

- **The Venezuelan city** of Merida has the world's highest cable car (altitude 4764m).

- **The world's highest waterfall** is the Angel Falls in Venezuela, plunging 979 m.

- **The Yanomami** are a native people who survive in remote forest regions of Venezuela and live by hunting with spears and gathering roots and fruit.

- **Kourou** in French Guiana is the launch site for European spacecraft such as the Ariane.

New York

> ★ STAR FACT ★
> New York is the USA's largest port and the finance centre of the world.

- **New York City** is the largest city in the USA and one of the largest in the world, with a population of 8 million.

- **Over 21 million people** live in the New York metropolitan area.

- **New York has five** boroughs: Manhattan, Brooklyn, the Bronx, Queens and Staten Island.

- **Manhattan** is the oldest part of the city, and is home to many attractions, including Central Park, Greenwich Village, the Rockefeller Center and Wall Street.

- **The 381 m high** Empire State Building, on Fifth Avenue, is one of the world's tallest and most famous buildings.

- **New York's most famous** statue is the Statue of Liberty, erected in 1886 at the entrance to New York harbour.

- **Dutch settler** Peter Minuit is said to have bought Manhattan island from the Iroquois Indians for trinkets with a value of $24.

- **New York** began in 1614 with the Dutch settlement of Fort Orange. It was renamed New York in 1664.

- **New York's famous** finance centre Wall Street is named after a protective wall built by Dutch colonists in 1653.

▼ The skyscrapers of Manhattan give New York one of the most famous skylines in the world.

Population

- **The world's population** climbed above 6 billion in 1999.

- **Over a quarter** of a million babies are born every day around the world.

- **World population** is growing at a rate of about 1.22% per year.

- **At the current rate** world population will hit 7.5 billion by 2020.

- **Between 1950** and 1990, the world's population doubled from about 2.5 billion to 5 billion, adding 2.5 billion people in 40 years.

- **The 1990s** added a billion people. The next decade will add 800 million. This adds 1.8 billion in 20 years.

Asia: 60.7%
Oceania: 0.5%
Europe: 12.4%
Antarctica: 0%
South America: 8.5%
Africa: 12.7%
North America: 5.2%

◄ *People are not spread evenly around the world. Some continents, like Europe, are densely populated. Antarctica is empty. The size of the figures in this diagram shows the size of the population of each continent. The size of the segment the figure is standing on shows the area of the continent.*

- **Asia has** about 60% of the world's population. China alone has 1.3 billion people and India has 1 billion.

- **The number of babies** born to each woman varies from 1.11 in Bulgaria to 7.11 in Somalia.

- **Latvia** has 100 women to every 86 men; Qatar has 184 men to every 100 women.

- **In the developed world** people are living longer. In Andorra people expect to live 83 years on average. In Mozambique, people only expect to live 36.5 years.

Berlin

- **Berlin** is Germany's capital and largest city, with a population of about 3.4 million.

- **Berlin** was originally capital of Prussia, which expanded to become Germany in the 1800s.

- **The city** was wrecked by Allied bombs in World War II.

- **After the War** Berlin was left inside the new communist East Germany and split into East and West by a high wall.

> ★ **STAR FACT** ★
> Almost every Berliner has a fragment of the Wall, torn down in 1989.

- **East Berlin** was the capital of East Germany; the West German capital moved to Bonn.

- **In 1989** the East German government collapsed and the Berlin Wall was torn down. East and West Germany were united in 1990 and Berlin was made capital again.

- **The Brandenburg Gate** is a huge stone arch built in 1791. It now marks the boundary between east and west.

- **Kurfurstendamm** is a famous shopping avenue. The Hansa quarter was designed by architects in the 1950s.

- **Since reunification** many spectacular new buildings have been built in Berlin including the refurbished Reichstag designed by Norman Foster.

◄ *When Germany was reunited, the old Reichstag became home of the German parliament again. It has now been given a major facelift.*

The USA

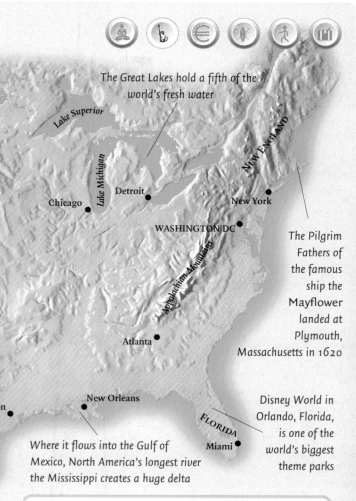

The Great Lakes hold a fifth of the world's fresh water

The Pilgrim Fathers of the famous ship the Mayflower landed at Plymouth, Massachusetts in 1620

Disney World in Orlando, Florida, is one of the world's biggest theme parks

Where it flows into the Gulf of Mexico, North America's longest river the Mississippi creates a huge delta

Map labels: Seattle, Cascade Range, Coast Ranges, Rocky Mountains, Lake Superior, NEW ENGLAND, Lake Michigan, Detroit, New York, Chicago, WASHINGTON DC, Great Basin, San Francisco, CALIFORNIA, Appalachian Mountains, Grand Canyon, Los Angeles, Atlanta, Houston, New Orleans, FLORIDA, Miami

▶ The United States is the richest and most powerful country in the world. Nearly 280 million people live here, and it covers a vast area of North America, from the freezing wastes of Alaska to the hot and steamy Everglades (marshes) of Florida.

- **Capital:** Washington DC. Area: 9,529,063 sq km. Currency: US dollar. Language: English.

- **Physical features:** Highest mountain: Mt McKinley (6194 m). Longest river: the Mississippi-Missouri-Red Rock (6020 km).

- **Population:** 281.4 million. Population density: 29/sq km. Life expectancy: men 73.4 years; women 80.1 years.

- **Wealth:** GDP: $8650 billion. GDP per head: $30,725.

- **Exports:** Road vehicles, chemicals, aircraft, generators, machinery, office equipment, scientific instruments.

- **Native Americans lived** in North America for 10,000 years before the Europeans arrived in the 16th century and gradually drove westwards, brushing the Native Americans aside. In 1788, English colonists founded the United States of America, now the world's oldest democratic republic, with a famous constitution (set of laws).

- **The USA** is the world's fourth largest in area, third largest in population, and has the largest GDP.

- **In the 1950s and 60s** Americans earned more money, ate more food, used more energy and drove more cars than anyone else in the world.

> ★ STAR FACT ★
> One in two Americans owns a computer –
> more than any other country in the world.

- **Now the USA** is the world's prime consumer of energy, oil, copper, lead, zinc, aluminium, corn, coffee and cocoa. It is also prime producer of aluminium and corn, and one of the top five producers of energy, oil, copper, lead, zinc, wheat and sugar.

▼ Through the films made in Hollywood, California, most of the world has become familiar with the American 'dream' of success.

Moscow and St Petersburg

▲ *St Petersburg is an elegant city with many beautiful houses and palaces such as the famous Hermitage museum.*

- **Moscow** is the largest city in the Russian Federation and capital of Russia.

- **Moscow** is Russia's main industrial centre, with huge textile and car-making plants, like the Likhachyov works.

> ★ STAR FACT ★
> Leningrad was dubbed 'Hero City' for its desperate defence against the Nazis from 1941-44.

- **Moscow's biggest shop** is Detsky Mir (Children's World).

- **Moscow's historic centre** is Red Square and the Kremlin, the walled city-within-a-city.

- **In the past** Moscow had wooden buildings and was often burnt down, most famously by Napoleon's troops in 1812.

- **Moscow is snow-covered** from November to April each year, but snow-ploughs keep all the main roads clear.

- **St Petersburg** is Russia's second largest city.

- **St Petersburg** was founded in 1703 by Tsar Peter the Great to be his capital instead of Moscow.

- **After the 1917 Russian Revolution**, communists called Petersburg (then called Petrograd) Leningrad and made Moscow capital. St Petersburg regained its name in 1991.

India

- **Capital:** New Delhi. Area: 3,287,263 sq km. Currency: Indian rupee. Languages: Hindi and English.

- **Physical features:** Highest mountain: K2 (8607 m). Longest river: Ganges (2510 km).

▼ *Hindu women in India traditionally wear beautifully coloured wraps or saris made of fine cloth such as silk.*

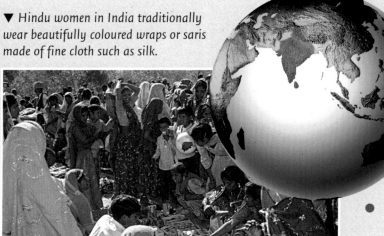

- **Population:** 1 billion. Population density: 310/sq km. Life expectancy: men 59.6 years; women 61.3 years.

- **Wealth:** GDP: $473.4 billion. GDP per head: $464.

- **Exports:** Gems, jewellery, clothes, cotton, fish, rice, textiles, engineering goods.

- **India has heavy** monsoon rains for six months of the year and dry weather for the rest.

- **India is the world's** largest democracy.

- **Two-thirds** of India's population grow their own food, mainly rice and wheat.

- **India's wheat production** has doubled since the Green Revolution of the 1960s when high-yield hybrids were introduced – but much wheat is sold abroad, pushing prices too high for many poor Indians.

- **India is the world's** 12th biggest industrial nation. Textiles remain important, but there is a growing emphasis on heavy industry, including iron and steel, vehicles, machine tools and pharmaceuticals.

Brazil

- **Capital:** Brasilia. Area: 8,547,404 sq km. Currency: Real. Language: Portuguese.

- **Physical features:** Highest mountain: Neblina (3014 m). Longest river: Amazon (6448 km).

- **Population:** 169.2 million. Population density: 19/sq km. Life expectancy: men 63.7 years; women 71.7 years.

- **Wealth:** GDP: $990 billion. GDP per head: $5845.

- **Exports:** Iron ore, coffee, timber, sugar, transport equipmet.

- **Brazil has one of** the biggest national debts of any country in the world – not far short of $200 billion.

Atlantic Ocean

Amazon River

Andes Mts

BRAZIL

Rio de Janeiro

Sao Paulo

Rio de la Plata

- ◄ Brazil is the world's fifth largest country, but most people live on the eastern edge. Much of the central area is cerrado (grass wilderness) or thick Amazon rainforest.

- **Brazil** is the world's biggest coffee grower. Soya, sugarcane, cotton, oranges, bananas and cocoa are also major crops.

- **The city of São Paulo** has grown faster than any other big city in the world and now 17.8 million people live there. Housing shortages in cities such as Rio de Janeiro and São Paulo mean 25 million Brazilians live in rickety sheds in sprawling shanty towns called *favelas*.

- **Brazilians are soccer-mad** and have won the World Cup more times than any other country.

- **The Amazon basin** contains the world's largest area of virgin rainforest – but an area almost the size of Ireland is being cleared each year for short-term cattle ranching.

North European food

- **Fish and bread** play a major role in the traditional Scandinavian diet.

- **Gravadlax** is a Swedish form of smoked salmon, usually served with pepper, dill and mustard sauce.

- **Smörgåsbord** is a Swedish speciality. It is a huge spread of bread and cold foods, including fish such as herring and salmon, and also cheeses.

- **Smörgåsbord** gets its name from the Swedish *smörgås*, meaning bread and *bord*, meaning table.

- **Every region in Germany** has its own range of foods, but things like *wurst* (sausages), pretzels and *sauerkraut* (pickled cabbage) are widely popular.

- **The German national drink** is beer, and every October a huge beer festival is held in Munich.

- **England is well known** for its hearty stews and winter roasts, especially roast beef. But the most popular food for those eating out is Indian.

- **An English speciality** is fish (deep-fried in batter)

and chips (fried slices of potato).

- **Vienna** in Austria is renowned for its coffee houses where the Viennese sit and eat *Kaffee und Kuchen* (coffee and cakes).

- **Poland is famous** for its rye bread and thick beet.

▼ The seas around Northern Europe were once teeming with fish and fish still plays a major role in the diet of people here.

Japan

- **Capital:** Tokyo. Area: 377,835 sq km. Currency: Yen. Language: Japanese.

- **Physical features:** Highest mountain: Mt Fujiyama (3776 m). Longest river: the Shinano-gawa (367 km).

- **Population:** 127 million. Population density: 337/sq km. Life expectancy: Men 77.6 years; women 84.2 years.

- **Wealth:** GDP: $4555 billion. GDP per head: $35,830.

- **Exports:** Electronic goods, steel, cars, ships, chemicals, textiles, machinery.

- **Japan** is very mountainous, so the big cities where nine out of ten people live are crowded into the coastal plains. 40 million people are crammed into Tokyo and its suburbs alone, making it the biggest urban centre in the world. Tokyo and the nearby cities have tall skyscrapers to make the most of the limited space available – but also with deep foundations, because Japan is prone to earthquakes.

- **Japan** is famous for its electronic goods – including walkmans and games consoles. It also makes huge amounts of steel, half the world's ships and more cars than any other country.

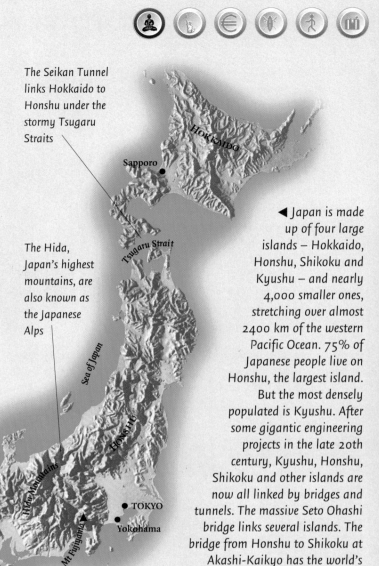

The Seikan Tunnel links Hokkaido to Honshu under the stormy Tsugaru Straits

The Hida, Japan's highest mountains, are also known as the Japanese Alps

◄ Japan is made up of four large islands – Hokkaido, Honshu, Shikoku and Kyushu – and nearly 4,000 smaller ones, stretching over almost 2400 km of the western Pacific Ocean. 75% of Japanese people live on Honshu, the largest island. But the most densely populated is Kyushu. After some gigantic engineering projects in the late 20th century, Kyushu, Honshu, Shikoku and other islands are now all linked by bridges and tunnels. The massive Seto Ohashi bridge links several islands. The bridge from Honshu to Shikoku at Akashi-Kaikyo has the world's longest single span – 2 km. Hokkaido and Honshu are linked by the Seikan tunnel, the world's longest undersea tunnel, 53.85 km long.

- **All but 14%** of the land is too steep for farming, but millions of little square rice fields are packed on to the coastal plains and on hillside terraces.

- **Most Japanese live** a very modern way of life. But traditions still survive and there are many ancient Buddhist and Shinto shrines.

▼ The beautiful, snow-capped Mt Fujiyama is the most famous of Japan's 1500 volcanoes and is sacred to the Shinto religion.

> ★ STAR FACT ★
> Japan has one of the world's largest fishing fleets which hauls in over 5 million metric tonnes of fish a year.

Australian landmarks

- **Australia's most famous landmark** is Uluru or Ayers Rock, the biggest monolith (single block of stone) in the world, 348 m high and 9 km around.
- **Uluru** is the tip of a huge bed of coarse sand laid down in an inland sea some 600 million years ago.
- **Lake Eyre** is Australia's lowest point, 15 m below sea level. It is also Australia's biggest lake by far, but it is normally dry and fills only once every 50 years or so.
- **Nullarbor plain** is a vast, dry plain in southern Australia. Its name comes from the Latin *nulla arbor* ('no tree').
- **Shark Bay** is famous for its sharks and dolphins.
- **Shark Bay** is also famous for its stromatolites, the world's oldest fossils, dating back 3.5 billion years. These are pizza-like mats made by colonies of blue-green algae.

> ★ STAR FACT ★
> The Great Barrier Reef is the world's largest structure made by living things.

▲ *Uluru is sacred to the Aboriginals. On its surface and in its caves are paintings made long ago by Aboriginal artists.*

- **The Murray-Darling River** is Australia's longest river (2739 km long).
- **The Great Barrier Reef** is a coral reef off the coast of Queensland in northwest Australia.
- **The Great Barrier Reef** is the world's biggest coral reef, over 2000 km long.

Ukraine and Belarus

- **Ukraine:** Capital: Kiev. Population: 50.8 million. Currency: Hryvnya. Language: Ukrainian.
- **Belarus:** Capital: Minsk. Population: 10 million. Currency: Belarusian rouble. Language: Belarusian.
- **Ukraine** is Europe's largest country (except for Russia), covering over 603,700 sq km.
- **Ukraine** is famous for its vast plains or *steppes*. The fertile black soils have made it 'the breadbasket of Europe', growing huge amounts of wheat and barley.
- **During the Soviet era** Soviet policies forced Ukrainians to speak Russian and adopt Russian culture, but the Ukrainian identity has been found again since they gained independence in 1991.

▶ *Ukraine and Belarus are flat countries that form the western margin of Russia, north of the Black Sea.*

- **In the Soviet era** over a quarter of Ukraine's industrial output was arms. Now Ukraine is trying to use these factories to make other products.
- **In 1986** a terrible accident occurred at the Chernobyl nuclear power plant north of Kiev. A reactor exploded spreading radioactivity over a wide area.
- **Nuclear energy** still provides 44% of Ukraine's power, but many Ukrainians are firmly against it.
- **Belarus** (known as Byelorussia under the USSR) is a flat country, covered in many places by thick forests and marshes. The Pripet Marshes are the largest in Europe covering 27,000 sq km.
- **Belarus** is known for making heavy-duty trucks, tractors and bicycles among other things. The forests provide products such as furniture, matches and paper.

Peoples of Australia

- **The Aborigines** make up 1.8% of Australia's population today, but they were the first inhabitants.

- **The word aborigine** comes from the Latin *ab origine*, which means 'from the start'.

- **Aborigine cave paintings** and tools have been found in Australia dating back to at least 45,000 years ago.

- **Aborigines** prefer to be called Kooris.

- **British and Irish people** began to settle in Australia about 200 years ago. They now form the majority of the population, along with other white Europeans.

- **Many of the earliest** settlers in Australia were convicts, transported from Britain for minor crimes.

◀ *The Kooris or Aborigines of Australia spread right across the Pacific many thousands of years ago and were probably the first inhabitants of America as well.*

- **Many Australians** have ancestral roots in the British Isles.

- **British and Irish settlers** drove the Aborigines from their land and 60% now live in cities.

- **After hard campaigning** some Aboriginal sacred sites are being returned to them, with their original names. Ayers Rock is now known as Uluru. A famous trial in 1992 returned to Aborigine Eddy Mabo land on Murray Island first occupied by his ancestors before the Europeans arrived.

- **Many recent immigrants** to Australia are from Southeast Asia, Serbia, Croatia and Greece.

Kazakhstan & neighbours

- **Kazakhstan:** Capital: Astana. Population: 16.9 million. Currency: Tenge. Language: Kazakh.

- **Uzbekistan:** Capital: Tashkent. Population: 25 million. Currency: Som. Language: Uzbek.

- **Turkmenistan:** Capital: Ashgabat. Population: 4.5 million. Currency: Manat. Language: Turkmen.

- **Some of the people** in this part of the world are still nomads, moving from place to place in search of new pastures for their herds.

- **Uzbekistan** has become wealthy from natural gas

◀ *The break-up of the Soviet Union left the countries around the Caspian Sea a legacy of some of the world's worst pollution. But they have a wealth of minerals and a venerable history.*

> ★ **STAR FACT** ★
> Kazakhstan has huge iron and coal reserves and the world's largest chrome mine.

and also from cotton, which they call 'white gold'.

- **The Baykonur Cosmodrome** in Kazakhstan is where the Russians launch most of their spacecraft.

- **The Soviet Union** forced nomads in Kalmykia by the Caspian Sea to boost sheep production beyond what the fragile steppe grass could handle. This created 1.4 million acres of desert.

- **The Aral Sea** on the Kazakh/Uzbek border was once the world's fourth largest lake. But irrigating farmland has cut the supply of water from the Amu Darya River and the Aral Sea is now shrinking rapidly.

- **The Caspian Sea** once had the sturgeon fish giving the most highly prized beluga caviar, but pollution has decimated the fish population.

East Africa

▲ *Tanzania is famous for safaris in the Serengeti park where lions, elephants, giraffes and many other animals are seen.*

- **Tanzania:** Legislative capital: Dodoma. Population: 33.7 million. Currency: Shilling. Languages: English, Swahili.

- **Rwanda:** Capital: Kigali. Population: 7.7 million. Currency: Franc. Languages: French, Kinyarwanda.

- **Burundi:** Capital: Bujumbura. Population: 7 million. Currency: Franc. Languages: French, Kirundi.

- **Uganda:** Capital: Kampala. Population: 22.2 million. Currency: Shilling. Languages: Swahili, English.

- **Malawi:** Capital: Lilongwe. Population: 10.9 million. Currency: Kwacha. Languages: Chichewa, English.

- **A fifth of Malawi** is taken up by Lake Nyasa, one of the world's largest, deepest lakes. .

- **The countries** of East Africa are the least urbanized in the world with 9 out of 10 living in the countryside.

- **In 1993 and 1994** Rwanda and Burundi were ravaged by one of the worst genocides in African history as tribal war flared between the Tutsi and Hutu peoples.

- **Lake Victoria** on the Tanzania and Uganda border covers 69,484 sq km, one of the world's largest lakes.

- **Tanzania's** main crops include maize, bananas andrice.

The Gran Chaco

- **The Gran Chaco** is a vast area of tropical grassland in Argentina, Paraguay and Bolivia.

- **It covers** an area of over 720,00 sq km, an area as large as northwest Europe.

- **It is home** to scattered native Indian groups such as the Guaycurú, Lengua, Mataco, Zamuco and Tupi-Guarani people.

- **The word Chaco** comes from the Quechua Indian word for 'Hunting Land' because it is rich in wildlife. *Gran* is Spanish for 'big'.

- **The major activities** on the Chaco are cattle grazing and cotton growing.

- **In the east** huge factories have been built to process tannin from the trees for leather production.

> ★ STAR FACT ★
> The sediments under the Gran Chaco are well over 3,000 m deep in places.

- **In places** grass can grow up to 3 m tall, higher than a rider on horseback.

- **The Chaco** is home to many wild animals, including pumas, tapir, rheas and giant armadillos.

- **The Chaco** is the last refuge of the South American maned or red wolf.

▶ *The jaguar is the Chaco's biggest hunting animal, and the biggest cat in the Americas. Yet unlike other big cats, it never roars. It just makes a strange cry rather like a loud sneeze.*

Poland and neighbours

▲ *Poland and its neighbours cluster around the Baltic Sea.*

- **Poland:** Capital: Warsaw. Population: 38.6 million. Currency: Zloty. Language: Polish.

- **Lithuania:** Capital: Vilnius. Population: 3.7 million. Currency: Litas. Language: Lithuanian.

- **Latvia:** Capital: Riga. Population: 2.4 million. Currency: Lats. Language: Latvian.

- **Estonia:** Capital: Tallinn. Population: 1.4 million. Currency: Kroon. Language: Estonian.

- **Poland** was led away from communism by trade union leader Lech Walesa, who became the first president democratically elected of Poland for 72 years.

- **The name Poland** comes from the Slavic word *polane* which means plain, and much of Poland is flat plains.

- **The shipyards at Gdansk** on the Baltic make Poland the world's fifth largest builder of merchant ships.

- **Krakow** has many historic buildings but the nearby Nowa Huta steelworks make it very polluted.

- **The Traditional way of life** in Latvia, Lithuania and Estonia suffered badly in the Soviet era. They are now rebuilding their identity.

- **Latvian** is one of the oldest European languages, related to the ancient Indian language Sanskrit.

West Coast USA

- **The western USA** is mountainous, with peaks in the Rockies, Cascades and Sierra Nevada soaring over 4,000 m.

- **Seattle** is the home of computer software giant Microsoft, and Boeing, the world's biggest aircraft maker.

- **Seattle** is the home of the Starbucks café chain – made famous by the TV series *Frasier*.

◄ *Sunset Boulevard in LA is a 32 km long road. Its Sunset Strip section is popular with film stars.*

- **Los Angeles** (LA) sprawls over a larger area than any other city in the world and has endless kilometres of freeways.

- **Film-makers** came to the LA suburb of Hollywood in 1908 because of California's sunshine. It has been the world's greatest film-making centre ever since.

- **The San Andreas fault** is the boundary between two huge continental plates. As it moves it gives west coast cities earthquakes. The worst may be yet to come.

- **San Francisco's** Golden Gate is named after the 1849 rush when prospectors came in thousands to look for gold.

- **California is** known as the 'Sunshine State'.

- **California's** San Joaquin valley is one of the world's major wine-growing regions.

> ★ STAR FACT ★
> Silicon Valley near San Francisco has the world's greatest concentration of electronics firms.

World religions

- **Christianity** is the world's largest religion, with 1.9 billion followers worldwide. Christians believe in a saviour, Jesus Christ, a Jew who lived in Palestine 2,000 years ago. Christ, they believe, was the Son of God. When crucified to death (nailed to a wooden cross), he rose from the dead to join God in heaven.

- **Islam** is the world's second largest religion with 1.3 billion believers. It was founded in Arabia in the 7th century by Mohammed, who Muslims believe was the last, greatest prophet sent by *Allah* (Arabic for God). The word *Islam* means 'act of resignation' and Muslims believe they must obey God totally and live by the holy book *The Koran*.

- **Hinduism** is almost 4,000 years old. Hindus worship many gods, but all believe in *dharma*, the right way to live. Like Buddhists, Hindus believe we all have past lives. By following the *dharma*, we may reach the perfect state of *Moksha* and so need never be born again.

- **Christianity** is split into three branches: Catholics whose leader is the Pope in Rome; Protestants; and the Eastern Orthodox church. Islam is split into Sunnis

and Shi'ites. Shi'ites are the majority in Iraq and Iran.

- **Buddhism** is the religion of 350 million SE Asians. It is based on the teachings of Prince Siddhartha Gautama, the Buddha, who lived in NE India from 563 to 483BC.

- **Judaism** is the religion of Jews. They were the first to believe in a single god, who they called *Yahweh*, over 4,000 years ago. There are over 10 million Jews living outside Israel and 4.4 million living in Israel.

- **Most of the world's** major religions except for Hinduism are monotheistic – that is, they believe in just one God.

- **Three million Muslims** visit their holy city of Mecca in Saudi Arabia every year on pilgrimage.

- **Jains** of India will not take any form of life. They eat neither meat nor fish, nor, usually, eggs. Jain priests often sweep paths in front of them as they go to avoid stepping on insects.

★ **STAR FACT** ★
The Hindu holy text, the *Bhagaavadgita*, contains almost 100,000 couplets. It is 7 times the length of the Iliad and Odyssey combined.

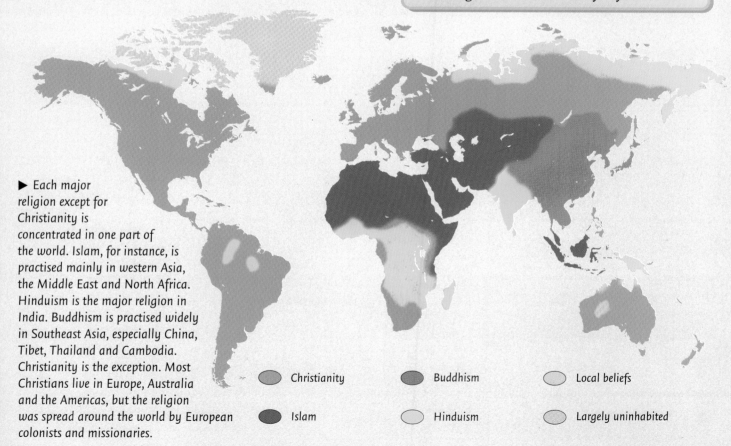

▶ Each major religion except for Christianity is concentrated in one part of the world. Islam, for instance, is practised mainly in western Asia, the Middle East and North Africa. Hinduism is the major religion in India. Buddhism is practised widely in Southeast Asia, especially China, Tibet, Thailand and Cambodia. Christianity is the exception. Most Christians live in Europe, Australia and the Americas, but the religion was spread around the world by European colonists and missionaries.

⬤ Christianity   ⬤ Buddhism   ⬤ Local beliefs

⬤ Islam   ⬤ Hinduism   ⬤ Largely uninhabited

Peoples of the Middle East

- **People have farmed** in the Middle East longer than anywhere else in the world.

- **The Middle East** was the site of the first cities and ancient civilizations such as those of Sumer and Babylon.

- **Most people** in the Middle East are Arabs.

- **Arabic is spoken** in all Middle East countries except for Iran where Farsi (Persian) is spoken, Turkey where most speak Turkish and Israel where most speak Hebrew.

- **Most people** in the Middle East are Muslims, but Lebanon has many Christians and Israel is mostly Jewish.

- **Many of the Arab** countries of the Middle East – except Israel – are dominated by Islamic traditions.

▲ *Many people in the Middle East wear traditional Arab head coverings.*

- **Islamic countries** of the Middle East are often ruled by kings and emirs, sultans and sheikhs who have absolute power. Yemen, Turkey and Israel are all republics. Iraq is a republic but it is ruled by President Saddam Hussein with absolute power.

- **The Jews of Israel** are locked in a conflict with the Arab people the roots of which which date back to the 1920s.

- **The people of the United Arab Emirates** (UAE) are among the richest in the world, with a yearly income of over $25,000 each.

- **The people of Yemen** are among the poorest in the world, with a yearly income of just $325 each.

Central America

- **Central American countries** are: Mexico, Guatemala, Belize, Honduras, El Salvador, Nicaragua, Costa Rica and Panama.

- **Mexico:** Capital: Mexico City. Population: 97 million. Currency: Peso. Language: Spanish.

- **Mexico City** is the world's second largest city after Tokyo, with a population of 18.4 million.

▼ *The Panama Canal cuts right across Central America to link the Atlantic and Pacific Oceans and save ships huge journeys.*

- **Most Central American** countries were torn apart by revolution and civil war in the 1900s, but are now quiet.

- **Mexico owes** in foreign debt almost $167 billion and pays over $37 billion a year back to other countries.

- **Much land** is used for 'cash crops' (crops that can be sold abroad for cash) like coffee rather than for food.

- **Many Central Americans** work the land, growing food for themselves or labouring on plantations.

- **Maize** (corn) has been grown in Mexico for 7,000 years to make things such as tortillas (cornflour pancakes).

- **Bananas** are the most important export in Central American countries, forming a third of Honduras's entire exports. While bananas are grown on lowlands, coffee beans are important exports for highland regions, especially in Nicaragua, Guatemala, Costa Rica and El Salvador.

- **Most of Mexico's people** are mestizos, descendants of both Spanish settlers and American Indians.

Georgia & its neighbours

◄ Georgia, Armenia and Azerbaijan lie in a band between the Black Sea and the Caspian Sea. Georgia and Armenia are mountainous. Azerbaijan contains flat plains.

● **Georgia:** Capital: Tbilisi. Population: 5.4 million. Currency: Lari. Language: Georgian.

● **Armenia:** Capital: Yerevan. Population: 3.7 million. Currency: Dram. Language: Armenian.

● **Azerbaijan:** Capital: Baku. Population: 8 million. Currency: Manat. Language: Azeri.

● **Georgia, Azerbaijan and Armenia** were once part of the Imperial Russia.

● **In Georgia more people** live to be 100 years old than anywhere else in the world except Japan.

● **Georgia's capital Tbilisi** is said to be one of the world's oldest cities.

● **The oil** under the Caspian Sea off Azerbaijan once helped Russia produce half the world's oil. Villages on floating platforms house oilworkers.

● **New oil strikes** suggest there is 200 billion barrels of oil under the Caspian Sea – as much as Iran and Iraq combined.

● **Oil has made** some people around the Caspian Sea rich, while others have remained desperately poor.

● **A 1520 km** pipeline from Kazakhstan's huge Tengiz field to Russia's Black Sea port of Novorossiysk has been opened.

South Africa

◄ Nelson Mandela was the hero of the struggle against apartheid in South Africa. In 1994 he became the country's first president elected by all the people.

● **Capitals:** Pretoria and Cape Town. Area: 1,219,080 sq km. Currency: Rand. Languages: 11 official languages including Zulu, Xhosa, English and Afrikaans.

● **Physical features:** Highest mountain: Injasuti (3408 m). Longest river: the Orange (2092 km).

● **Population:** 43.7 million. Population density: 36/sq km. Life expectancy: men 47.3 years; women 49.7 years.

> ★ STAR FACT ★
> In the 1900s, almost half the world's gold came from South Africa.

● **Wealth:** GDP: $150.3 billion. GDP per head: $3440.

● **Exports:** Gold, diamonds, pearls, metals, metal products, machinery, citrus fruit, wine.

● **Until 1991** people of different races in South Africa were separated by law. This was called apartheid.

● **Apartheid** meant many black people were forced to live in specially built townships such as Soweto. Townships are far from cities and workplaces, so workers must commute for hours each on crowded buses.

● **South Africa** has two capital cities. The administration is in Pretoria and parliament is in Cape Town.

● **The Kruger National Park** supports the greatest variety of wildlife species on the African continent.

Germany

▲ Neuschwanstein, built for 'Mad' King Ludwig II of Bavaria in the 1870s, is the most famous of the many castles in Bavaria and Germany's Rhineland.

► The flatter northern part of Germany is a mixture of heath, marsh and rich farmland, where cereals such as rye are widely grown. The south is mountainous, with powerful rivers flowing through deep, wooded valleys.

Hamburg is Germany's biggest port

Germany's northwest is known as Lower Saxony

North Sea

Baltic Sea

Hamburg

North German Plain

BERLIN

RUHR • Dortmund
• Düsseldorf
• Cologne
• Bonn

Dresden

• Frankfurt

When trees began to die from acid rain in the famous Black Forest, many Germans became committed to the Green cause. The country now has strong environment protection laws

• Stuttgart

BAVARIA

Black Forest

• Munich

Zugspitze

Alps

Germany's longest river, the Danube rises in the Black Forest in Germany. It empties into the Black Sea

- **Capital:** Berlin. Area: 356,973 sq km.
  Currency: euro.
  Language: German.

- **Physical features:** Highest mountain: Zugspitze (2962 m). Longest river: Danube (2850 km).

- **Population:** 82.7 million. Population density: 231/sq km. Life expectancy: men 74 years; women 80.3 years.

- **Wealth:** GDP: $2257 billion. GDP per head: $27,300.

- **Exports:** Machinery, vehicles, chemicals, iron, steel, textiles, food, wine.

- **Germany** is the world's third biggest industrial nation after the USA and Japan, famous for its precision engineering and quality products, such as tools and machine tools.

- **Germany's smoky industrial** heartland was the Ruhr valley, where dozens of coal mines fed huge steelworks. Many mines and steelworks have now closed and many

people have moved south to places like Stuttgart and Munich to escape unemployment and dirty air. But the Ruhr remains important to industry.

- **Germany is** the world's third biggest car-maker after the USA and Japan. It is well known for its upmarket cars such as Mercedes, BMW and Audi.

- **German farms** are often small, family-run affairs. Yet the country can grow almost all its own food – growing huge quantities of cereal and sugar beet, and raising large numbers of cows and pigs.

★ STAR FACT ★
The reunification of Germany in 1990 made it western Europe's biggest country by far.

North Africa

- **Morocco:** Capital: Rabat. Population: 29 million. Currency: Dirham. Language: Arabic.
- **Algeria:** Capital: Algiers. Population: 31.6 million. Currency: Algerian dinar. Language: Arabic.
- **Tunisia:** Capital: Tunis. Population: 9.8 million. Currency: Tunisian dinar. Language: Arabic.
- **Libya:** Capitals: Tripoli and Surt. Population: 6.4 million. Currency: Libyan dinar. Language: Arabic.
- **Much of** the world's phosphate supply comes from Morocco and Tunisia.
- **Algeria and Libya** both have large reserves of oil and gas.
- **People in Morocco,** Tunisia and Algeria eat a lot of couscous. Couscous is made from wheat which is pounded into hard grains of semolina, then steamed until soft. The couscous is then served with stewed lamb or vegetables.
- **The Moroccan** custom is to eat using the right hand rather than knives and forks.
- **The historic cities** of Fez and Marrakesh in Morocco are famous for their colourful souks or markets, where thousands of tourists each year come to haggle over beautiful hand-woven carpets, leather goods and jewellery.

> ★ STAR FACT ★
> Libya is building the 3870 km Great Manmade River, the world's longest water pipe, to irrigate 800sq km of land.

▶ The coastal areas of northwest Africa have warm, Mediterranean climates and farmers grow things like olives and citrus fruits.

Pacific food

- **Most places** around the Pacific are near the sea, so fish plays an important part in diets.
- **In Japan** fish is often eaten raw in thin slices called sashimi, or cooked with vegetables in batter as a dish called tempura, often served with soy sauce.
- **At home** most Japanese eat traditional foods including rice and noodles, as well as fish, tofu and vegetables or eggs.
- **When out,** many Japanese people eat American-style foods from fast food restaurants.

▶ Lightly grilled or barbecued giant prawns and other seafood play a major role in Pacific food.

- **The Japanese** eat only half as much rice now as they did in 1960, as younger people prefer bread and doughnuts.
- **Younger Japanese** people have a diet richer in protein and fat than their parents', so grow 8-10 cm taller.
- **Pacific islanders** traditionally ate fish like bonito and tuna and native plants like breadfruit, coconuts, sweet potatoes and taro. They made flour from sago palm pith.
- **Many islanders** now eat mainly canned Western food and suffer malnutrition.
- **Filippino** food is a mix of Chinese, Malay, American and Spanish. *Adobo* is chicken or pork in soy sauce.
- **Some Australians** now often eat 'fusion' food which blends Asian with European cooking styles.

London

▲ London's Houses of Parliament and its tower with its bell Big Ben were built in 1858 after a fire destroyed an earlier building.

- **London** is the capital of the United Kingdom and its largest city by far, with a population of about 7.2 million.

- **People have settled** here for thousands of years, but the city of London began with the Roman city of Londinium.

- **Throughout the 19th century** London was the world's biggest city, with a million people, and the hub of the world's largest empire, the British Empire.

- **London** is based on two ancient cities: the City of London, which developed from the Roman and Saxon towns, and Westminster, which developed around the palaces of English kings around 1,000 years ago.

- **London** has 500,000 factory workers, but most people work in services, such as publishing and other media. London is one of the world's major finance centres.

- **Eight million tourists** come to London each year.

- **London's tallest building** is 244 m Canary Wharf tower.

- **The London Eye** is the biggest wheel in the world, giving people a bird's eye view over London.

- **London's oldest large buildings** are the Tower of London and Westminster Abbey, both 1,000 years old. The Tower of London was built for William the Conqueror.

> ★ **STAR FACT** ★
> 700,000 people work in banking and finance – more than in any other city in the world.

Iraq and Iran

- **Iran:** Capital: Tehran. Population: 76.4 million. Currency: Iranian rial. Language: Persian (Farsi).

- **Iraq:** Capital: Baghdad. Population: 23.1 million. Currency: Iraqi dinar. Language: Arabic.

- **Iran is the largest** non-Arabic country in the Middle East.

- **Iran was once** called Persia, and was the centre of an empire ruled by the Shah that dates back thousands of years. The last Shah was overthrown in 1979.

- **Iran is an Islamic** country, and the strong views of religious leader Ayatollah Khomeini (who died in 1989) played a key role in the revolution in 1979, which brought him to power.

▶ Iraq and Iran are mostly hot, dry countries, but Iran is much more mountainous, ringed by the Zagros and Elburz mountains.

- **Iran is famous for its carpets,** often called Persian carpets. They are Iran's second largest export, after oil. Oil brings Iran 80% of its export earnings.

- **Iraq** was the place where civilization probably began 7,000 years ago. The Greeks called it Mesopotamia.

- **Since 1979** Iraq has been ruled by Saddam Hussein and his leadership has brought the country into conflict with much of the world – especially when he invaded Kuwait and started the Gulf War in 1991 when the USA and other nations retaliated.

- **Only about** a sixth of Iraq is suitable for farming and so it has to import much of its food, but it is one of the world's major oil producers.

- **United Nations** sanctions applied after the Gulf War still restrict trade with Iraq. Some argue that it is poor Iraqis who suffer from these and not Saddam Hussein.

West Africa

- **West African countries** are: Cape Verde, Liberia, Equatorial Guinea, Niger, Mauritania, Mali, Burkina Faso, Senegal, Gambia, Guinea Bissau, Liberia, Guinea, Ivory Coast, Ghana, Sierra Leone, Togo, Benin and São Tomé and Prîncipe. All have populations under 10 million except for Ghana, Ivory Coast and Mali.

- **Ghana:** Capital: Accra. Population: 18.4 million. Currency: Cedi. Language: English.

- **Ivory Coast:** Capital: Yamoussoukro. Population: 15.1 million. Currency: CFA franc. Language: French.

- **Mali:** Capital: Bamako. Population: 12.6 million. Currency: CFA franc. Language: French.

- **West Africa** grows over half the world's cocoa beans.

- **Yams** are a vital part of the diet of people in West Africa, often providing breakfast, dinner and tea.

- **West Africa** is rich in gold and diamonds, which once sustained ancient Mali and its capital of Timbuktu.

▲ Ivory Coast alone grows 40% of the world's cocoa beans. Here beans are drying in the sun.

- **Ghana and Guinea** are rich in bauxite (aluminium ore).

- **Ghana** was called Gold Coast by Europeans because of the gold used by the Ashanti peoples there.

- **Ghana** is still poor, but many of its young people are the best educated in Africa.

Peoples of North America

- **82%** of the population of North America are white descendants of immigrants from Europe.

- **Among the smaller groups** 13% are black, 3% are Asian and 1% are American Indians.

- **Hispanics** are descended from a mix of white, black and American Indian people from Spanish-speaking countries of Latin America such as Mexico, Puerto Rico and Cuba. 12% of the US population is Hispanic.

- **92%** of the population of the USA was born there. Many new immigrants are Hispanic.

- **The original peoples** of North America were the American Indians who were living here for thousands of years before Europeans arrived.

▲ The original peoples of North America were the Indians, but they have been overwhelmed by European settlers.

- **The native people** of America were called Indians by the explorer Christopher Columbus, but they have no collective name for themselves. Most American Indians prefer to be identified by tribe.

- **There are about** 540 tribes in the USA. The largest are the Cherokee, Navajo, Chippewa, Sioux. and Choctaw

- **Most black Americans** are descendants of Africans brought here as slaves from 1600 to 1860.

- **Most European** immigrants before 1820 were from Britain, so the main language is English.

- **Spanish** is spoken by many Americans and French is spoken by 24% of Canadians.

The Amazon

- **The Amazon River** in South America is the world's second longest river (6448 km), and carries far more water than any other river.

- **The Amazon basin** – the area drained by the Amazon and its tributaries – covers over 7 million sq km and contains the world's largest tropical rainforest.

- **Temperatures** in the Amazon rainforest stay about 27°C all year round.

- **The Amazon rainforest** contains more species of plant and animal than anywhere else in the world.

- **The Amazon is home** to over 60,000 different plants, 1550 kinds of bird, and 3000 species of fish in its rivers.

- **Manaus** in the Amazon basin has a population of over a million and a famous 19th century opera house.

> ★ **STAR FACT** ★
> The Amazon basin is home to more than
> 2 million different kinds of insect.

▲ *In its upper reaches in the Andes, the Amazon tumbles over 5000 m in the first 1000 km.*

- **Since the 1960s** the Brazilian government has been building highways and airports in the forest.

- **10%** of the forest has been lost for ever as trees are cut for wood, or to clear the way for gold-mining and ranching.

- **Forest** can sometimes regrow, but has far fewer species.

Egypt and neighbours

- **Egypt:** Capital: Cairo. Population: 68.1 million. Currency: Egyptian pound. Language: Arabic.

- **Ethiopia:** Capital: Addis Ababa. Population: 66.2 million. Currency: birr. Language: Amharic, Oromo.

- **Sudan:** Capital: Khartoum. Population: 29.8 million. Currency: Sudanese dinar. Language: Arabic.

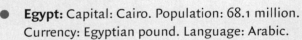

- **Egypt relies** heavily on tourists visiting its ancient sites, like the 4700 year-old Great Pyramid of Cheops at Giza.

- **99% of Egyptians** live by the River Nile which provides water for farming, industry and drinking. A vast reservoir, Lake Nasser, was created when the Nile was dammed by the Aswan High Dam.

- **Cairo** has a population of 11.6 million and is growing so rapidly there are major housing and traffic problems.

- **Lots of cotton** is grown in Egypt and the Sudan.

- **Sudan is the largest** country in Africa.

- **In the 1980s and 1990s** the people of Ethiopia, Sudan and Somalia suffered dreadful famine. Many people here are still very poor and without enough to eat.

- **Grasslands south of the Sahara** are dotted with acacia thorn trees which ooze a liquid called gum arabic when their bark is cut. This was used in medicine and inks.

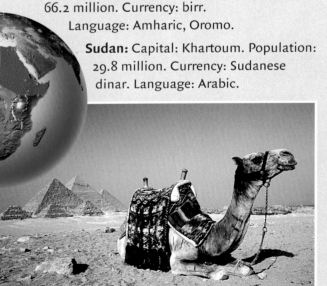

◄ *Camels have provided desert transport in Egypt for thousands of years. The pyramids of the pharaohs are in the background.*

Australia

- **Capital:** Canberra. Area: 7,682,300 sq km. Currency: Australian dollar. Language: English.

- **Physical features:** Highest mountain: Mt Kosciuszko (2230 m). Longest river: Murray-Darling (3750 km).

- **Population:** 18.8 million. Population density: 2/sq km. Life expectancy: men 76.8 years; women 82.2 years.

- **Wealth:** GDP: $428.7 billion. GDP per head: $22,755.

- **Exports:** Ores and minerals, coal, oil, machinery, gold, diamonds, meat, textiles, cereals.

- **Australia** enjoyed its own goldrush when gold was discovered there in 1851.

- **Most of Australia** is so dry only 6% is good for growing crops, although the country sells a lot of wheat.

But huge areas are used for rearing cattle and sheep, many raised on vast farms called 'stations'. Australia is also famous for its wines.

- **Australia has huge amounts** of iron, aluminium, zinc, gold and silver. The Mount Goldsworthy mine in Western Australia alone is thought to have 15 billion tonnes of iron ore. Cannington in Queensland is the world's largest silver mine.

- **Australia's climate** encourages outdoor activities like surfing. Thousands head for Bondi Beach near Sydney on Christmas Day for a party or to surf. Australia is also the world's top cricketing nation.

Darwin
Arnhem Land
Gulf of Carpentaria
Great Barrier Reef
Kimberley Plateau
Great Sandy Desert
QUEENSLAND
Alice Springs
Great Artesian Basin
Brisbane
Lake Eyre
WESTERN AUSTRALIA
Flinders Ranges
Great Australian Bight
Perth
Adelaide
Sydney
CANBERRA
Mt Kosciusko
Melbourne
Bass Strait
TASMANIA
Hobart

The railtrack across Nullabor Plain is the world's longest straight track

▲ Australia is the world's smallest continent but sixth biggest country. Much of it is dry and thinly populated. Most people live in the southeast or along the coast.

The Great Dividing Range divides the moist coastal plain from the dry outback

★ STAR FACT ★
Australia's 115 million sheep produce more than a quarter of the world's wool.

Grand Canyon

- **The Grand Canyon** in Arizona in the southwest USA is one of the most spectacular gorges in the world.

- **The Grand Canyon** is about 466 km long and and varies in width from less than 1 km to over 30 km.

- **In places** the Grand Canyon is so narrow that motorcycle stunt riders have leaped right across from one side to the other.

- **The Grand Canyon** is about 1600 m deep, with almost sheer cliff sides in some places.

- **Temperatures** in the bottom of the Canyon can be as much as 14°C hotter than they are at the top, and the bottom of the Canyon gets only 180 mm of rain per year compared to 660 mm at the top.

▲ *The shadows cast by the evening sun reveal the layer upon layer of rock in the steep sides of the Grand Canyon.*

- **The Grand Canyon** was cut by the Colorado River over millions of years as the whole Colorado Plateau was rising bit by bit. The bends in the river's course were shaped when it still flowed over the flat plateau on top, then the river kept its shape as it cut down through the rising plateau.

- **As the Colorado** cut down, it revealed layers of limestone, sandstone, shale and other rocks in the cliffs.

- **The Colorado** is one of the major US rivers, 2334 km long.

- **The Hoover Dam** across the Colorado is one of the world's highest concrete dams, 221 m high.

- **The Hoover Dam** creates 185 km long Lake Mead, North America's biggest artificial lake.

Peru and neighbours

- **Peru:** Capital: Lima. Area: 1,285,216 sq km. Currency: Nuevo Sol. Languages: Spanish and Quechua.

- **Physical features:** Highest mountain: Huascaran (6768 m). Longest river: Amazon (6448 km).

- **Population:** 25.7 million. Population density: 20/sq km. Life expectancy: men 65.6 years; women 69.1 years.

- **Wealth:** GDP: $70.2 billion. GDP per head: $2730.

- **Exports:** Fish products, gold, copper, zinc, iron, oil, coffee, llama and alpaca wool, cotton.

- **Peru** is the third largest country in South America. The coastal plain is desert, but Peru's biggest city Lima is here. Inland are the towering Andes mountains, where rivers have cut deep gorges.

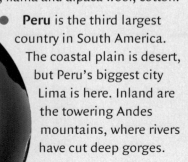

- **Peru** was the home of the Inca Empire conquered by the Spaniard Francisco Pizarro in the 1520s. Now it has a larger Indian population than any other South American nation.

- **Peru** is a leading producer of copper, lead, silver and zinc, and a major fishing nation. But most people are poor, especially in the mountains. In the 1990s guerillas called *Sendero Luminoso* (Shining Path) and *Tupac Amaru* sparked off violent troubles.

- **Ecuador:** Capital: Quito. Population: 12.7 million. Currency: US dollar. Language: Spanish.

- **Bolivia:** Capital: La Paz. Population: 8.3 million. Currency: boliviano Language: Spanish.

▶ *The llama was for centuries the main source of meat and wool and the main means of transport for people in Peru.*

Kenya

- **Capital:** Nairobi. Area: 582,646 sq km. Currency: Kenyan shilling. Languages: Swahili and English.

- **Physical features:** Highest mountain: Mt Kenya (5199 m). Longest river: Tana (708 km).

- **Population:** 30.3 million (est). Population density: 52/sq km. Life expectancy: men 47.3 years; women 48.1 years.

- **Wealth:** GDP: $11.5 billion. GDP per head: $380.

- **Exports:** Tea, coffee, fruit, flowers, vegetables, petroleum products.

- **Remains of early human ancestors** found by Lake Turkana show people have lived in Kenya for millions of years.

- **Much of Kenya** is a vast, dry, grassland plain, home to spectacular wildlife such as lions, giraffes and elephants. They attract thousands of tourists each year.

- **Most Kenyans** live on small farm settlements, raising crops and livestock for themselves, but there are big cash crop plantations for tea, coffee, vegetables and flowers.

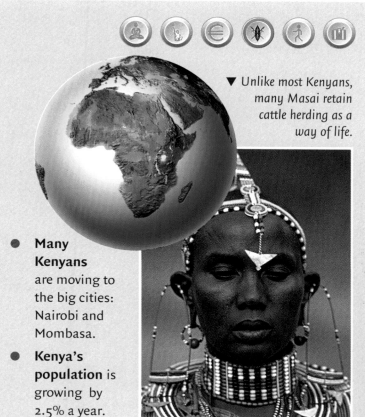

▼ Unlike most Kenyans, many Masai retain cattle herding as a way of life.

- **Many Kenyans** are moving to the big cities: Nairobi and Mombasa.

- **Kenya's population** is growing by 2.5% a year.

Health and education

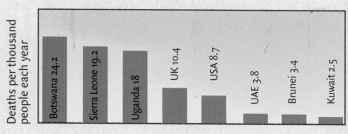

- **Progress** in medical science, better diet and improved hygiene have made the world a healthier place for many.

- **How long** people are likely to live is called life expectancy. In 1950, the world average was just 40 years. Now it is over 63 years.

- **Life expectancy** is usually high in richer countries. The Andorrans live on average for 83.5 years; the Japanese live for 80.8 years.

- **Life expectancy** is much lower in poor countries. People in Zambia live just 37.3 years; people in Mozambique live 36.5 years.

- **Vaccination programmes** have reduced the effects of some major diseases. The terrible disease smallpox was thought to be wiped out in 1977.

- **Some diseases** are on the increase in poorer parts of the world. AIDS (Acquired Immune Deficiency Syndrome) is now killing huge numbers of Africans.

- **In some parts** of the world, disease, lack of food and water and poor healthcare mean that one child in every four dies before reaching the age of five in poor countries like Afghanistan and Sierra Leone.

- **In the USA and Europe** less than one child in a hundred dies before the age of five.

- **In wealthier** countries such as Italy and Switzerland, there is on average one doctor for every 350 people.

- **In most poor African** countries, there is just one doctor for every 50,000 people.

Deaths per thousand people each year

| Botswana 24.2 | Sierra Leone 19.2 | Uganda 18 | UK 10.4 | USA 8.7 | UAE 3.8 | Brunei 3.4 | Kuwait 2.5 |

▲ Death rates per thousand people vary from over 20 in many African countries to under 3 in many Arab countries of the Gulf.

France

> ★ **STAR FACT** ★
> French vineyards produce more than 22% of
> the world's wine every year.

- **Capital:** Paris. **Area:** 543,965 sq km. **Currency:** euro.
  Language: French.

- **Physical features:** Highest mountain: Mont Blanc
  (4810 m). Longest river: Loire (1020 km).

- **Population:** 58.7 million. Population density:108/sq
  km. Life expectancy: men 74.9 years; women 83.6 years.

- **Wealth:** GDP: $1407 billion. GDP per head: $24,330.

- **Exports:** Agricultural products, chemicals,
  machinery, vehicles, pharmaceuticals.

- **France** is the biggest food producer in Europe,
  apart from Russia. In the north and west,
  wheat, sugar beet and many other crops
  are grown and dairy cattle are raised. In
  the warmer, drier south of France
  grapes and other
  fruit are grown.

▶ *France is famous
both for its beef
cattle, like this
Charolais, and its
dairy cows, which
give cheeses
including the
delicious soft cheeses
Brie and Camembert.*

- **France has limited** coal and oil reserves, but
  nuclear power gives France 75% of its energy.

- **France** is the biggest country in western Europe.
  Much is still rural, with ancient farmhouses and
  villages looking as if they have changed little in
  centuries. But French cities like Lyon and Marseille
  are famous for their sophisticated culture. They are
  also the centres of so much industry that
  France is the world's fifth largest
  industrial nation after the USA,
  Japan, Germany and UK.

- **The French**
  are traditionally
  famous for their
  haute cuisine (fine
  cooking) using
  the best
  ingredients, and
  creating fantastic
  table displays.
  Later, plainer styles
  developed and a lighter
  style of cooking called
  nouvelle cuisine (new
  cooking) has
  developed to suit
  today's tastes.

▶ *France
is a large and
enormously varied
country. In the
centre are the rugged
hills and volcanic peaks of
the Massif Central. To the
south is the warm sunny
Mediterraean coast. The
low, rolling countryside of
the north and west is
cooler. The highest
mountains are the Alps in
the southeast and the
Pyrenees in the
southwest, along the
border with Spain.*

Calais
Lille
English Channel
Le Havre
PARIS
CHAMPAGNE
Strasbourg
BRITTANY
Le Mans
Bay of Biscay
Tours
Avallon
La Châtre
La Rochelle
Lyon
Mont Blanc ▲
Bordeaux
Massif Central
Bergerac
Jura
LANGUEDOC
MONACO
PROVENCE
Nice
Toulouse
Marseille
Mediterranean Sea
Pyrenees

Antarctica

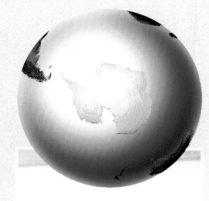

- **Antarctica** is the fifth largest continent, larger than Europe and Australia, but 98% of it is under ice.

- **The Antarctic population** is made up mostly of scientists, pilots and other specialists there to do research in the unique polar environment.

- **About 3000 people** live in Antarctica in the summer, but less than 500 stay all through the bitter winter.

- **The biggest community** in Antarctica is McMurdo which is home to up to 1200 people in summer and has cafés, a cinema, a church and a nuclear power station.

- **People and supplies** reach McMurdo either on ice-breaker ships that smash through the sea ice, or by air.

- **McMurdo settlement** was built around the hut the British polar explorer Captain Scott put up on his 1902 expedition to the South Pole.

- **The Amundsen–Scott** base is located directly underneath the South Pole.

- **Antarctica** has a few valuable mineral resources including copper and chrome ores.

- **There is coal** beneath the Transarctic Mountains, and oil under the Ross Sea.

- **Under the Antarctic Treaty** of 1961, 27 countries agreed a ban on mining to keep the Antarctic unspoiled. They allow only scientific research.

◀ *Emperor penguins are among the few large creatures that can survive the bitter Antarctic winter. They breed on the ice cap itself.*

Industry

▲ *Nearly every country in the world is becoming more and more industrialized. New industries include services like banking rather than traditional manufacturing.*

- **Primary industries** are based on collecting natural resources. They include farming, forestry, fishing, mining and quarrying.

- **Things made** by primary industries are called primary products or raw materials.

- **Primary industries** dominate the economies of poorer countries. Copper is 80% of Zambia's exports.

- **Primary products** are much less important in developed countries. Primary products earn 2% of Japan's GDP.

- **Secondary industry** is taking raw materials and turning them into products from knives and forks to jumbo jets. This is manufacturing and processing.

- **Tertiary industries** are the service industries that provide a service, such as banking or tourism, not a product.

- **Tertiary industry** has grown enormously in the most developed countries, while manufacturing has shrunk.

- **'Postindustrialization'** means developing service industries in place of factories.

- **Tertiary industries** include internet businesses.

> ★ **STAR FACT** ★
> More than 70% of the UK's income now comes from tertiary industry.

The Middle East

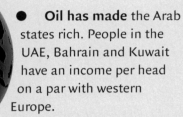

- **Saudi Arabia:** Capital: Riyadh. Population: 21.7 million. Currency: Rial. Language: Arabic.

- **Yemen:** Capital: Sana. Population: 18.1 million. Currency: Riyal. Language: Arabic.

- **Kuwait:** Capital: Kuwait City. Pop: 2.3 m. Curr: dinar. Language: Arabic.

- **United Arab Emirates (UAE):** Capital: Abu Dhabi. Population: 2.4 million. Currency: Dirham. Language: Arabic.

- **Population:** Oman: 2.7 million. Bahrain: 620,000. Qatar: 590,000.

- **Much of the Middle East** is desert. Rub'al Khali in Saudi Arabia lives up to its name, 'Empty Quarter'. Nomads called Bedouins have herded sheep and goats here for thousands of years. Now most Bedouins live in houses.

- **Oil has made** the Arab states rich. People in the UAE, Bahrain and Kuwait have an income per head on a par with western Europe.

- **Saudi Arabia** is the world's leading exporter of oil and second only to Russia as oil producer. It has 25% of the world's known oil reserves.

- **Yemen** is one of the world's poorest countries.

- **The oil-rich states** along the Gulf are short of water. Most comes from wells, but now they are building desalination plants which remove salt so they can use water from the sea.

◀ *Civilization began in the Middle East, but the climate dried and turned much of it to desert.*

The Russian steppes

- **The steppes** are a vast expanse of temperate grassland, stretching right across Asia.

- **'Steppes'** is a word meaning lowland.

- **The Western Steppe** extends 4000 km from the grassy plains of the Ukraine through Russia and Kazakhstan to the Altai mountains on the Mongolian border.

- **The steppes** extend 300-800 km from north to south.

★ STAR FACT ★
The steppes extend 8000 km across Eurasia, a fifth of the way round the world.

- **The Eastern Steppe** extends 2500 km from the Altai across Mongolia to Manchuria in north China.

- **The Eastern Steppe** is higher and colder than the Western Steppe and the difference between winter and summer is as extreme as anywhere on Earth.

- **Nomadic herders** have lived on the steppes for over 6000 years.

- **It was on the steppes** near the Black and Caspian Seas that people probably first rode horses 5000 years ago.

- **The openness** of the steppes meant that travel by horse was easy long before roads were built.

◀ *The steppes have supported nomadic herding people for thousands of years, but their way of life is rapidly dying out.*

American food

▲ The American hamburger has been spread around the world by fast-food chains. The American people eat 45,000 hamburgers every minute!

- **Many American** foods were brought from Europe by immigrants.

- **Hamburgers** were brought to the USA by German immigrants in the 1880s, but are now the most famous American food.

- **Frankfurters** came from Frankfurt in Germany (though this is disputed by people from Coburg, now in Bavaria). They became known in the USA as 'hot dogs' by the early 1890s.

- **The pizza** came from Naples in Italy, but the first pizzeria opened in New York in 1895. Pizzas caught on after 1945.

- **The bagel** originated in Poland early in the 17th century where it was known as beygls. It was taken to New York by Jewish immigrants and is often eaten filled with smoked salmon and cream cheese.

- **Self-service** cafeterias began in the 1849 San Francisco Gold Rush.

- **The world's first** fast-food restaurant may have been the White Castle which opened in Wichita, Kansas in 1921.

- **The world's biggest** fast-food chain is McDonalds which has over 29,000 branches worldwide.

- **Pies** have been popular in the US since colonial times, and apple pie is the symbol of American home cooking.

- **American home cooking** includes beef steaks, chicken and ham with potatoes plus a salad. But Americans eat out often – not only fast-food such as hamburgers and French fries, but Chinese, Italian and Mexican dishes.

Southern Africa

- **Mozambique:** Capital: Maputo. Population: 19.6 million. Currency: Metical. Language: Portuguese.

- **Angola:** Capital: Luanda. Population: 12.8 million. Currency: Kwanza. Language: Portuguese.

- **Zambia:** Capital: Lusaka. Pop: 9.9 million. Currency: Kwacha. Language: English.

- **Population:** Namibia: 1.7 million. Botswana: 1.6 million. Swaziland: 985,000.

- **Large areas** of southern African countries are too dry to farm intensively. Most people grow crops such as maize or raise cattle to feed themselves.

- **In Mozambique** plantations grow crops like tea and coffee for export, but most people who work on them are poor.

- **In 2000** much of Mozambique was devastated by huge floods from the Zambezi and Limpopo Rivers.

- **Zambia is the** world's fourth largest copper producer and relies on copper for 80% of its export earnings.

- **Namibia is the** world's second largest lead producer.

- **Namibia** has the world's biggest uranium mine and an estimated three billion carats of diamond deposits.

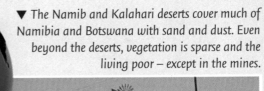

▼ The Namib and Kalahari deserts cover much of Namibia and Botswana with sand and dust. Even beyond the deserts, vegetation is sparse and the living poor – except in the mines.

Tokyo

- **Tokyo** is the world's biggest city. With the port of Yokohama and the cities of Chiba and Kawasaki, it makes an urban area that is home to 27.3 million people.

- **Tokyo** is Japan's capital and leading industrial and financial centre.

- **Tokyo's stock exchange** is one of the world's three giants, along with London and New York.

- **Tokyo** was originally called Edo when it first developed as a military centre for the Shoguns. It was named Tokyo in 1868 when it became imperial capital.

- **14,000 people** live in every square kilometre of Tokyo – twice as many as in the same area in New York.

- **Some hotels** in Tokyo have stacks of sleeping cubicles little bigger than a large refrigerator.

> ★ STAR FACT ★
> Tokyo probably has more neon signs than any other city in the world.

▲ *Tokyo is perhaps the busiest, most crowded city in the world.*

- **During rush hours** *osiyas* (pushers) cram people on to commuter trains crowded with 10 million travellers a day.

- **Traffic police** wear breathing apparatus to cope with traffic fumes.

- **Tokyo mixes** the latest western-style technology and culture with traditional Japanese ways.

Canada

- **Capital:** Ottawa. Area: 9,970,610 sq km. Currency: Canadian dollar. Languages: English, French.

- **Physical features:** Highest mountain: Mt Logan (5951 m). Longest river: the Mackenzie, linked to the Peace by the Great Slave Lake (4241 km).

- **Population:** 30.7 million. Population density: 3/sq km. Life expectancy: men 76.2 years; women 81.9 years.

- **Wealth:** GDP: $683.6 billion. GDP per head: $22,280.

- **Exports:** Vehicles and parts, machinery, petroleum, aluminium, timber, wood pulp, wheat.

- **Canada** is the world's second largest country.

- **Three-quarters** of Canada's small population live within 100 km of the southern border with the USA; the rest of the country is rugged and only thinly inhabited.

- **Only 5%** of Canada is farmed, but this is a big area. The Prairie provinces – Saskatchewan, Alberta and Manitoba – grow a lot of wheat and raise many cattle.

- **Canada** has 10% of the world's forest and is the world's largest exporter of wood products and paper.

- **The Inuit people** of the far north in the Arctic were given their own homeland of Nunavut in 1999.

◄ *The completion of the Canadian Pacific railroad right across Canada in 1886 was one of the great engineering feats of the 1800s.*

Energy

- **Humans** now use well over 100 times as much energy as they did 200 years ago, and the amount is rising.

- **Europe, North America and Japan** use 70% of the world's energy with just a quarter of the people.

- **Fossil fuels** are coal, oil and natural gas – fuels made from organic remains buried and fossilized over millions of years. Fossil fuels provide 90% of the world's energy.

- **Fossil fuel** pollutes the atmosphere as it burns, causing health problems, acid rain and also global warming.

- **Fossil fuel** is non-renewable. This means it can't be used again once burned. At today's rates, the world's coal and oil will be burned in 70 years and natural gas in 220 years.

- **Renewable energy** like running water, waves, wind and sunlight will not run out. Nuclear energy is non-renewable, but uses far less fuel than fossil fuel.

- **Alternative energy** is energy from sources other than fossil fuels and nuclear power. It should be renewable and clean.

- **Major alternative energy** sources are waves, geothermal, tides, wind and hydro-electric power.

Food for living

Home

Industry

Transport

◀▼ Each person in developed countries uses 10 times as much energy as each person in less developed countries.

Energy use in developed countries

Energy use in less developed countries

- **The Sun** provides the Earth with about the same as 500 trillion barrels of oil in energy a year – 1,000 times as much as the world's oil reserves. Yet little is used. Solar panels provide just 0.01% of human energy needs.

▼ The pie diagram in the centre shows how much of the world's energy is provided by different sources. The top layer shows proportions ten years ago. The bottom layer shows proportions now. See how biomass energy use has risen.

Oil is our most important energy source, now providing almost 40% of the world's energy needs. The biggest reserves are in the Middle East and Central Asia.

Coal still provides almost 23% of the world's energy needs. Two-thirds of the world's reserves are in China, Russia and the USA. India and Australia are major producers too.

Wood and dried animal dung – called biomass – provide the main fuel for half the world's population. In some poorer countries, it provides 90% of all fuel.

Natural gas provides over 22% of world energy needs, and the proportion is rising. The biggest reserves are in Russia, the USA and Canada.

Hydrolectric power (HEP) uses fast-flowing rivers or water flowing through a dam to generate electricity. HEP supplies 7% of world energy needs.

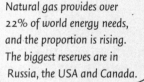

Nuclear power now provides about 5% of the world's energy needs. The major producers are France, the USA and Russia.

Geothermal power uses heat from deep inside the Earth – either to heat water or make steam to generate electricity. Experts think geothermal use will go up.

Windpower, wavepower and solar energy produce barely 5% of the world's energy needs. The proportion is going up, but only very, very slowly.

Peoples of southern Asia

▶ Thai people are descended from peoples who migrated from China between the 11th and 12th centuries.

- **There is a huge** variety of people living in southern Asia from India to the Philippines.

- **India** has hundreds of different ethnic groups speaking 30 languages and 1652 dialects.

- **Indonesia** also has many different ethnic groups and over 400 different languages and dialects.

- **In Cambodia,** Vietnam, Thailand, Myanmar and Sri Lanka, people are mostly Buddhists.

- **In Indonesia,** Malaysia, Pakistan and Bangladesh, people are mostly Muslim.

- **In India,** 81% of people are Hindus.

- **The word Hindu** comes from the Indus river where Dravidian people created one of the world's great ancient civilizations 4,500 years ago.

- **By Hindu tradition** people are born into social classes called castes. Members of each caste can only do certain jobs, wear certain clothes and eat certain food.

- **Most Indians** are descended from both the Dravidians and from the Aryans who invaded and pushed the Dravidians into the south about 3,500 years ago.

- **The people of East Timor** in Indonesia are mainly Christian. Before they became independent they were under the oppressive rule of the Indonesian military government.

Afghanistan & neighbours

- **Afghanistan:** Capital: Kabul. Population: 25.6 million. Currency: Afghani. Language: Pashto, Dari.

- **Tajikistan:** Capital: Dushanbe. Population: 6.4 million. Currency: Tajik rouble. Language: Tajik.

- **Kyrgyzstan:** Capital: Bishkek. Population: 4.5 million. Currency: Som. Language: Kyrgyz.

- **Nepal:** Capital: Kathmandu. Population: 24.4 million. Currency: Nepalese rupee. Language: Nepali.

- **These four** countries plus Tibet contain most of the world's highest mountains, including Everest (8863 m) and Kanchenjunga (8598 m) in Nepal and Garmo in Tajikistan (7495 m).

- **The blue** gemstone lapis lazuli has been mined at Sar-e-Sang in Afghanistan for over 6000 years.

- **Kyrgyzstan** has been independent from the USSR since 1991, but 16% of the people are Russian or half-Russian. Only a few still live in the traditional kyrgyz tents or 'yurts'.

- **Tajikistan** is still mostly rural and farmers in the deep valleys grow cotton and melons.

- **Afghanistan** was wracked by war for 17 years until the fiercely Muslim Taliban came to power in the late 1996. Their regime was overthrown in 2001.

▼ The Himalayas, highlighted on these maps, are the world's highest mountains.

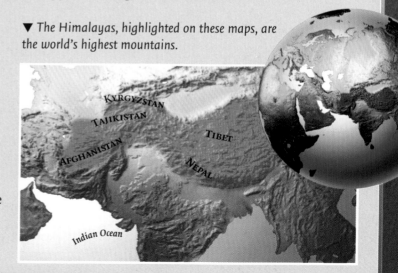

Hungary and neighbours

▲ Hungary is one of the world's leading producers of sunflower oil, and in summer vast areas of its Great Plain turn yellow with sunflower blooms.

- **Hungary:** Capital: Budapest. Population: 10 million. Currency: Forint. Language: Hungarian.
- **Czech Republic:** Capital: Prague. Population: 10.2 million. Currency: Koruna. Language: Czech.

- **Slovakia:** Capital: Bratislava. Population: 5.3 million. Currency: Koruna. Language: Slovak.
- **Slovenia:** Capital: Ljubljana. Population: 2 million. Currency: Tolar. Language: Slovenian.
- **Until 1990** all of these countries were under Soviet rule except Slovenia. It was part of Yugoslavia.
  - **Since 1990**, the historic city of Prague has become a popular destination, especially with the young, and many recording artists have worked in studios here.
  - **The Czech Republic** is famous for Pilsen beer brewed in the town of Plzen with hops grown locally.
  - **Hungary's** national dish is *goulash*. This is a rich stew made from meat, onion and potatoes, spiced with paprika (red pepper) and served with black rye bread.
- **Slovakian people** were largely rural, with a strong tradition of folk music, dancing and dress. Now over half the population has moved into industrial towns.
- **Vienna's white** Lippizaner horses are bred in Slovenia.

Hong Kong

> **! NEWS FLASH !**
> The proposed Landmark tower in Kowloon could be 576 m tall.

- **Hong Kong** is a Special Adinistrative Region on the coast of China. It comprises a peninsula and 237 islands.
- **Hong Kong** was administered by the British from 1842 until July 1 1997.
- **6.8 million** people are crowded into Hong Kong, mostly in the cities of Kowloon and Hong Kong itself.
- **Hong Kong** is one of the world's most bustling, dynamic, overcrowded cities. It makes huge amounts of textiles, clothing and other goods and is also one of the world's major financial and trading centres.
- **All but 3%** of Hong Kong people are Chinese, but many speak English as well as Chinese.
- **Hong Kong** is one of the world's three biggest ports, along with Rotterdam and Singapore.

- **Hong Kong** is the world's biggest container port.
- **Hong Kong's Chep Lap Kok** airport, opened in 1998, is one of the world's most modern airports.
- **The Hong Kong-Shanghai Bank** tower is one of the world's most spectacular modern office blocks.

▼ Hong Kong is quite mountainous, so people are crowded on to a small area of land, often in huge, high-rise apartment blocks.

China

★ STAR FACT ★

A baby is born in China every two seconds, and someone dies every two and a half.

- **Capital:** Beijing. Area: 9,573,998 sq km. Currency: Yuan. Language: Guoyo (Mandarin).

- **Physical features:** Highest mountain: Qomolanjma Feng (Mt Everest, 8863 m). Longest river: Chang Jiang (Yangtze) (6300 km).

- **Population:** 1276.3 million. Population density: 133/sq km. Life expectancy: men 68.1 years; women 71.1 years.

- **Wealth:** GDP: $1392 billion. GDP per head: $1000.

- **Exports:** Electrical machinery, textiles and clothing, footwear, toys and games, iron and steel, crude oil, coal, tobacco.

- **China** is the world's third largest country, stretching from the soaring Himalayas in the west to the great plains of the Huang (Yellow) and Chang Jiang (Yangtze) rivers in the east where most people live.

- **China** is the most highly populated country in the world. In 1979, it was growing so rapidly the government made it illegal for couples to have more than one child. In the countryside, people disliked the law because extra children were needed to work the fields. In towns, the law worked better, but single children were often spoiled and so called 'Little Emperors'.

- **68%** of China's people still live and work on the land, growing rice and other crops to feed themselves. But as China opens up to western trade, so industry is growing in cities like Guangzhou (Canton) and more and more country people are going to work there.

- **China** became communist in 1949, and the nationalist government fled to the island of Taiwan. China and Taiwan still disagree over who should govern China. Taiwan now has its own thriving economy and makes more computer parts – especially microchips – than any other country in the world.

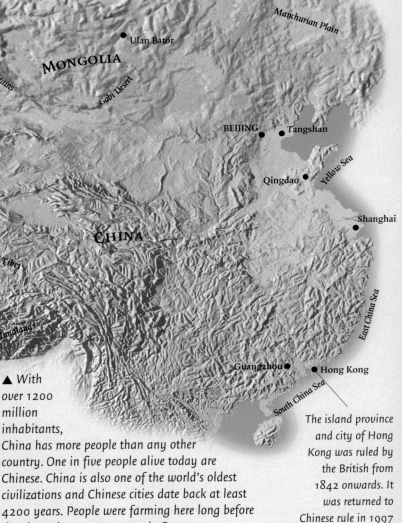

▲ Rice is grown all over China in flooded fields called paddies.

▲ With over 1200 million inhabitants, China has more people than any other country. One in five people alive today are Chinese. China is also one of the world's oldest civilizations and Chinese cities date back at least 4200 years. People were farming here long before the pharaohs came to power in Egypt.

The island province and city of Hong Kong was ruled by the British from 1842 onwards. It was returned to Chinese rule in 1997

Netherlands and Belgium

- **Netherlands:** Capitals: The Hague and Amsterdam. Population: 15.9 million. Currency: euro. Language: Dutch.

- **Belgium:** Capital: Brussels. Population: 10.2 million. Currency: euro. Languages: Flemish, French.

- **Belgium and the Netherlands** along with Luxembourg are often called the Low Countries because they are quite flat. The Netherlands' highest hill is just 321 m.

- **A quarter of the Netherlands** (also known as Holland) is polders – land once covered by the sea, but now protected by banks called dykes and pumped dry.

- **The Netherlands** exports more cheese than any other country in the world. Edam and Gouda are famous.

- **The Netherlands** is famous for its vast fields of tulips.

> ★ **STAR FACT** ★
> The Netherlands is the world's biggest trader in cut flowers.

- **Rotterdam** at the mouth of the Rhine is one of the world's biggest ports, handling a million tonnes of goods each day.

- **Brussels** is the seat of the European Union Commission and Council.

- **The Belgian** city of Antwerp is the diamond-cutting centre of the world.

▶ Holland is famous for its windmills. These are not for grinding flour but for working the pumps that keep the flat land dry.

Peoples of South America

- **South America** has a population of a little over 345 million people.

- **Before its conquest** by the Spanish and Portuguese in the 16th century, South America was home to many native peoples.

- **There are native villages** in the Andes with only one race, and a few native tribes in the Amazon rainforest who have had little contact with the outside world.

- **The main population groups now** are American Indians, whites, blacks (whose ancestors were brought as slaves) and people of mixed race.

▶ In the Amazon, small tribes like the Matses still survive as they have done for thousands of years.

- **Most people** in Latin America are mixed race.

- **The largest mixed race** groups are mestizos (people with both American Indian and white ancestors) and mulattoes (people with black and white ancestors).

- **Mestizos** are the majority in countries such as Paraguay and Venezuela. Mulattoes are the majority in Brazil.

- **The Europeans** who came to South America were mostly Spanish and Portuguese, so nearly two-thirds of South Americans speak Spanish.

- **Many American Indians** speak their own languages.

- **Quechua** is a native language, which Peru has made its official language along with Spanish.

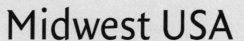

Midwest USA

- **Huge amounts** of wheat and maize are grown on the damper eastern side of the vast rolling plains.
- **Millions** of beef cattle are raised on ranches in the drier west.
- **The weather is often extreme** here with scorching summer days and winter blizzards.
- **Tornado Alley** is a band through Kansas and beyond which is blasted by hundreds of tornadoes every summer.
- **Heavy farming** in the 1930s let dry winds strip away soil leaving just dust over a vast area called the Dust Bowl. Irrigation and windbreaks have lessened the problem.
- **Millions of buffalo (bison)** roamed the Great Plains 200 years ago. Now just 50,000 live on reserves.

◄ The midwest is North America's agricultural heartland, raising millions of cattle and growing vast areas of yellow corn.

- **Detroit** on the Great Lakes is the centre of the US car and truck industry. Ford, Chrysler and General Motors all have their headquarters here.
- **Detroit** is sometimes known as Motown (short for 'motor town') and was famous in the 1960s for its black soul 'Motown' music.
- **Many Italians** have emigrated to the USA and most US cities have an Italian area. Chicago's Italians invented their own deep, soft version of the pizza.
- **Chicago,** known as 'The Windy City', is the USA's third largest city, home to over nine million people.

Ireland

- **Capital:** Dublin. Area: 70,285 sq km. Currency: euro. Languages: Irish and English. Over two-thirds of Ireland is in the Republic of Ireland. The rest is Northern Ireland, which is part of the UK.
- **Physical features:** Highest mountain: Carrauntoohil (1041 m). Longest river: Shannon (372 km).
- **Population (Republic):** 3.7 million. Population density: 53/sq km. Life expectancy: men 73.8 years; women 79.4 years.
- **Wealth (Republic):** GDP: $99.8 billion. GDP per head: $26,880.
- **Exports:** Machinery, transport equipment, chemicals.
- **Peat** is one of Ireland's few natural energy resources. Peat is the compressed rotten remains of plants found in peat bogs. Once dried it can be burned as fuel.
- **Ireland** is famous for its pubs, folk music and hospitality. Large numbers of Irish are returning home because of its booming economy and labour shortages.
- **Ireland,** and Dublin in particular, have been thriving in recent years – partly because of the success of high-tech electronics and computer industries.
- **Entertainment** is big business in Ireland. Many films – both American and Irish – are made in Ireland, and pop, rock and folk music are huge money earners.

◄ Right out on the northwest of Europe in the Atlantic, Ireland is a mild, moist place. Damp mists and frequent showers keep grass lush and green, earning the island the name 'The Emerald Isle'.

> ★ STAR FACT ★
> Ireland has enjoyed the fastest growing economy in Europe since the mid 1990s.

North and South Korea

- **North Korea:** Capital: Pyongyang. Population: 22.2 million. Currency: Won. Language: Korean.

- **South Korea:** Capital: Seoul. Population: 46.1 million. Currency: Won. Language: Korean.

- **Korea** split into the communist North and capitalist South in 1948.

- **A bitter war** between North and South involving the USA ended with a treaty in 1953.

- **North and South Korea** both still have large armies. South Korea's has over half a million soldiers.

- **After the war** money from US banks helped the South, for a time, to become the world's fastest growing economy.

- **Huge factories** run by companies called chaebol churned out everything from computers to Hyundai and Daewoo cars.

- **South Korean shipyards** build one in six of the world's ships. Only Japan builds more.

- **Since 1997** North Korea has suffered food shortages after two years of floods followed by drought.

- **In 1997-98** the uncovering of massive government dishonesty made many South Korean businesses bankrupt. New President Kim Dae Jung has led a recovery.

◄ The Korean company of Daewoo not only makes cars, but ships, computers, TVs, videos and much more besides.

Central Africa

- **Equatorial Guinea:** Capital: Malabo. Population: 450,000. Currency: CFA franc. Languages: Fang, Spanish.

- **Gabon:** Capital: Libreville. Population: 1.2 million. Currency: CFA franc. Language: French.

- **Congo (Brazzaville):** Capital: Brazzaville. Pop: 3 million. Currency: CFA franc. Languages: Monokutuba, French.

- **Cameroon:** Capital: Yaoundé. Population: 15.1 million. Currency: CFA Franc. Languages: Fang, French, English.

- **Central African Republic (CAR):** Capital: Bangui. Population: 3.6 million. Currency: CFA Franc. Languages: Sango, French.

- **French and English** are official languages, but most people speak their own African language.

- **Most Gabonese** are farmers, but Gabon is rich in oil, iron and manganese and famous for its ebony and mahogany.

- **70% of people** in Cameroon are farmers, growing crops such as cassava, corn, millet, yams and sweet potatoes. Most of Cameroon's roads are dust roads.

- **Congo (Brazzaville)** is so called to distinguish it from neighbouring DR Congo. It is one of the world's poorest countries. It is thickly forested and many people travel by dugout canoes. Many raise bananas or grow crops to feed themselves.

- **The CAR and Equatorial Guinea** are among the least developed countries in Africa.

▶ Cameroon, Gabon, CAR, Congo and Equatorial Guinea lie near the Equator, and are often thickly wooded with rainforest.

Nigeria and neighbours

- **Nigeria:** Capital: Abuja. Population: 128.8 million. Currency: Naira. Languages: include English Creole, Hausa, Yoruba, English.

- **Niger:** Capital: Niamey. Population: 10.8 million. Currency: CFA franc. Languages: Hausa, French.

- **Chad:** Capital: N'Djamena. Population: 7.3 million. Currency: CFA franc. Languages: Arabic and French.

- **Most people** in the north of Nigeria, and in Niger and Chad live by growing food for themselves.

- **The amount of rainfall** increases dramatically from north to south, and the vegetation changes from rainforest to dry grassland to desert in marked bands.

- **In the dry north** people grow mainly millet; in the moist south, they grow rice and roots such as cassava and yam.

- **Oil makes up 95%** of Nigeria's exports.

- **The money from oil** has made Nigeria among the most heavily urbanized and populous countries in Africa, especially around its main city Lagos.

- **Nigeria** is home to over 250 different peoples.

- **Nigeria** became a democracy again in 1998 after years of bitter civil war and military dictatorship, but tensions remain.

▼ *The position of Nigeria, Niger and Chad on the southern fringes of the Sahara desert makes them prone to drought as climate change and overgrazing push the desert further south.*

Argentina

▼ *'Gaucho' means orphan, but the tough gauchos are Argentina's heroes.*

- **Capital:** Buenos Aires. Area: 2,766,889 sq km. Currency: Peso. Language: Spanish.

- **Physical Features:** Highest mountain: Aconcagua (6960 m). Longest river: Paraná (4880 km).

- **Population:** 37 million. Population density: 13/sq km. Life expectancy: men 69.7 years; women 76.8 years.

- **Wealth:** GDP: $367 billion. GDP per head: $9900.

- **Exports:** Wheat, maize, meat, hides, wool, tannin, linseed oil, peanuts, processed foods, minerals.

- **The Argentinian landscape** is dominated by the pampas, the vast flat grasslands which stretch all the way to the high Andes mountains in the west.

- **Most of Argentina's** exports are pampas products – wheat, corn, meat, hides and wool.

- **Cattle** on the pampas – 49 million of them – are herded by Argentina's famous cowboys, the gauchos.

- **Argentina** is the most educated country in South America, with a third of students going to university.

★ **STAR FACT** ★
Argentineans eat more meat than any other nation in the world.

The United Kingdom

- **Capital:** London. Area: 244,088 sq km. Currency: Pound. Language: English.

- **Physical features:** Highest mountain: Ben Nevis (1343 m). Longest river: Severn (354 km).

- **Population:** 59.7 million. Population density: 245/sq km. Life expectancy: men 74.5 years; women 79.8 years.

- **Wealth:** GDP: $1450 billion. GDP per head: $24,280.

- **Exports:** Manufactured goods such as chemicals and electronics, financial services, music and publishing.

- **The British Isles** are 4,000 islands with 20,000 km of coast. There are two large islands: Great Britain and Ireland. The United Kingdom (UK) is four countries joined politically – England, Scotland, Wales and N. Ireland.

- **England** is intensively farmed, especially in the south where wheat, barley, rape, sugar beet and vegetables are grown. In the moister west and north of England, especially, and Scotland and Wales, cattle and sheep are reared.

- **The Industrial Revolution** began in the UK. In the 1800s, heavy industries such as steelmaking and engineering grew in northern coalfield cities like Manchester and Leeds. Now coal's importance as an energy source has dwindled, and some smaller northern towns are finding it difficult to survive. But southern England is thriving on light industries and services.

- **London** is one of the world's great financial centres. Over 500 international banks are crammed into a small area of the city called the Square Mile. Here billions of dollars' worth of money deals are done every day.

▲ The Millennium Dome was erected by the River Thames in east London to celebrate the year 2000.

◀ England is the biggest and most densely populated of the countries of the UK – a lush land of rolling hills, rich farmland and big cities. Wales is a land of hills and sheep farms, except for the south where industry is important and coal was once mined in huge amounts. Much of Scotland is wild moors and valleys, and most people live in the central lowlands around the cities of Glasgow and Edinburgh. A third of Northern Ireland's population lives in Belfast.

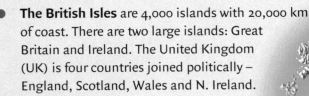

SCOTLAND
Ben Nevis
Grampian Mountains
Atlantic Ocean
EDINBURGH
Glasgow
Southern Uplands
N. IRELAND
BELFAST
North Sea
Irish Sea
IRELAND
DUBLIN
Manchester
Liverpool
Cambrian Mts.
St. George's Channel
Cork
WALES
Birmingham
ENGLAND
CARDIFF
Bristol
LONDON
Dover
Plymouth
English Channel

★ STAR FACT ★
Over 60% of the UK's workforce now work in financial and service industries.

The Near East

- **Syria:** Capital: Damascus. Population: 16.1 million. Currency: Syrian pound. Language: Arabic.
- **Jordan:** Capital: Amman. Population: 5.2 million. Currency: Jordan dinar. Language: Arabic.
- **Lebanon:** Capital: Beirut. Population: 3.3 million. Currency: Lebanese pound. Language: Arabic.
- **Damascus** was a major trading centre 4000 years ago.
- **Syria** is at the western end of the belt of rich farmland known as the fertile crescent, which was the cradle of the earliest civilizations, along the banks of the Tigris and Euphrates rivers.
- **Most Syrian farmers** still work on small plots growing cotton and wheat. But 40% of Syrians now work in services.
- **Around 70% of Jordan's** income is from services like tourism and banking.

- **The people of Syria,** Jordan and Lebanon are mostly Arabs. 86% of Syrians and 96% of Jordanians are Muslims, but 35% of Lebanese are Christians.
- **In 1948** Palestine was split between Israel, Jordan and Egypt. Palestinian Arabs' desire for their own country has caused conflict with Israelis.
- **In 1996** Israeli troops withdrew from the Gaza strip region, and Palestinians elected their own local administration.

Indian food

- **Most Indians** live on very plain diets – based on staples such as rice in the east and south, *chapatis* (flat wheat bread) in the north and northwest, and *bajra* (millet bread) in the Maharashtra region.
- **The staple foods** are supplemented by *dal* (lentil porridge), vegetables and yoghurt.
- **Chillis and other spices** such as coriander, cumin, ginger and turmeric add flavour.
- **Chicken and mutton** are costly and eaten occasionally. Hindus will not eat beef and Muslims will not eat pork.
- **Many Indian meals** are cooked in *ghee* (liquid butter). Ghee is made by heating butter to boil off water, then allowing it to cool and separate. Ghee is scooped off the top.

- **Although many Indians** have simple diets, India has an ancient and varied tradition of fine cooking.
- **Curries are** dishes made with a sauce including the basic Indian spices – turmeric, cumin, coriander and red pepper. The word curry comes from the Tamil *kari*, or sauce.
- **The basis of a curry** is a *masala*, a mix of spices, often blended with water or vinegar to make a paste.
- **Southern Indian** vegetable curries are seasoned with hot blends like *sambar podi*.
- **Classic northern Indian** Mughal dishes are often lamb, or chicken based, and seasoned with milder *garam masala*.

◀ *An Indian meal is rarely served on a single plate. Instead, it comes in different dishes, which diners dip into.*

Paris

▲ Les Halles was the main market for Paris from the 12th century, but in the 1970s was transformed into a modern shopping centre.

- **Paris** is the capital of France and its largest city with a population of over nine million.

- **Paris** is France's main business and financial centre. The Paris region is also a major manufacturing region, notably for cars.

- **Paris is famed** for luxuries like perfume and fashion.

- **Paris is known** for restaurants like La Marée, cafés like Deux Magots and nightclubs like the Moulin Rouge.

- **Paris monuments** include the Arc de Triomphe, Eiffel Tower, Notre Dame cathedral and the Beauborg Centre.

- **Paris gets its name** from a Celtic tribe called the Parisii who lived there 2000 years ago.

- **The Roman general** Julius Caesar said the Parisii were 'clever, inventive and given to quarrelling among themselves'. Some say this is true of Parisians today.

- **Paris was redeveloped** in the 1850s and 60s by Baron Haussman on the orders of Emperor Napoleon III.

- **Haussman** gave Paris broad, tree-lined streets called boulevards, and grand, grey, seven-storey houses.

> ★ **STAR FACT** ★
> Well over half of France's business deals are done in Paris.

Balkan peninsula

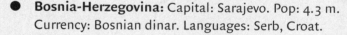

- **The Balkan peninsula** is a mountainous region in SE Europe between the Adriatic and Aegean Seas.

- **The Balkans** include several different nations, some of which were under either the Austro-Hungarian or Turkish Empires until 1918.

- **In 1945 Yugoslavia** became six republics in a federal Communist state. This ended in 1991.

- **Bosnia-Herzegovina,** Croatia, Macedonia, Slovenia and Kosovo broke away from Yugoslavia in the 1990s amid much bitter conflict.

- **Serbia and Montenegro** formed a smaller Yugoslavia in 1992. This ceased to exist in May 2002 when the country was reformed as Serbia and Montenegro.

- **Serbia and Montenegro:** Capital: Belgrade. Population: 10.5 million. Currency: Dinar, euro. Language: Serb.

- **Croatia:** Capital: Zagreb. Population: 4.5 million. Currency: Kuna. Language: Croat.

- **Bosnia-Herzegovina:** Capital: Sarajevo. Pop: 4.3 m. Currency: Bosnian dinar. Languages: Serb, Croat.

- **Albania:** Capital: Tirana. Population: 3.5 million. Currency: New lek. Language: Albanian.

- **Macedonia:** Capital: Skopje. Pop: 2.2 m. Curr: Dinar. Languages: Macedonian, Albanian.

▼ This bridge at Mostar in Bosnia-Herzegovina was a casualty of the wars of the 1990s.

Agriculture

▶ Most of the world's food is grown in the Northern Hemisphere or Asia. Asia is a major grower of wheat, rice, sweet potatoes, sorghum and all pulses such as beans. In fact, 90% of all rice and sweet potatoes are grown in Asia. Half the world's corn is grown in North America. 40% of potatoes are grown in Europe.

Millet

Sweet Potato

Barley

Potato

Rice

Maize

Casava

Wheat

Oats

Soya Bean

★ STAR FACT ★
There are now over twice as many farm animals in the world as humans – over 14 billion.

- **Only 12%** of the Earth's ice-free land surface is suitable for growing crops – that is, about 13 billion hectares. The rest is either too wet, too dry, too cold or too steep. Or the soil is too shallow or poor in nutrients.

- **A much higher** proportion of Europe has fertile soil (36%) than any other continent. About 31% is cultivated.

- **In North America** 22% of the land is fertile but only 13% is cultivated, partly because much fertile land is lost under concrete. Surprisingly, 16% of Africa is potentially fertile, yet only 6% is cultivated.

- **Southern Asia** is so crowded that even though less than 20% of the land is fertile, over 24% is cultivated.

- **Dairy farms** produce milk, butter and cheese from cows in green pastures in fairly moist parts of the world.

- **Mixed farming** involves both crops and livestock as in the USA's Corn Belt, where farmers grow corn to feed pigs and cattle. Many European farms are mixed, too.

- **Mediterranean farming** is in areas with mild, moist winters and warm, dry summers – like California, parts of South Africa and the Mediterranean. Here winter crops include wheat, barley and broccoli. Summer crops include peaches, citrus fruits, grapes and olives.

- **Shifting cultivation** involves growing crops like corn, rice, manioc, yams and millet in one place for a short while, then moving on before the soil loses goodness.

- **Shifting cultivation** is practised in the forests of Latin America, in Africa and parts of Southeast Asia.

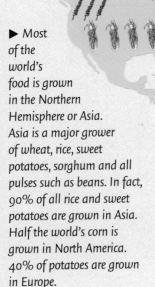

◀ In places farming is now a highly mechanized industry, but in SE Asia many farmers work the land as they have for thousands of years.

The West Indies

▶ Famous Jamaica rum is made from cane sugar, still the West Indies' major crop despite the rise of beet sugar.

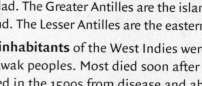

- **Cuba:** Capital: Havana. Population: 11.2 million. Currency: Cuban peso. Language: Spanish.

- **Jamaica:** Capital: Kingston. Population: 2.6 million. Currency: Jamaican dollar. Language: English.

- **The four largest islands** in the West Indies are Cuba, Hispaniola, Jamaica and Puerto Rico. Hispaniola is split into two countries: Haiti and the Dominican Republic.

- **The islands** are mostly in a long curve stretching from Cuba to Trinidad. The Greater Antilles are the islands of the western end. The Lesser Antilles are the eastern end.

- **The original inhabitants** of the West Indies were Carib and Arawak peoples. Most died soon after the Spanish arrived in the 1500s from disease and abuse.

- **Today most West Indians** are descended from Africans brought here as slaves to work on the sugar plantations.

- **The slaves** were freed in the mid-1800s, but most people here are still poor and work for low wages.

- **In Haiti** only one person in 250 has a car; fewer than 1 in 10 has a phone.

- **Many people** work the land on sugar, banana or coffee plantations, and also farm a plot to grow their own food.

- **Many tourists** come for the warm weather and clear blue seas.

Vietnam and neighbours

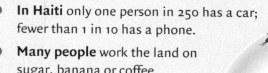

- **Vietnam:** Capital: Hanoi. Population: 80.5 million. Currency: dong. Language: Vietnamese.

- **Laos:** Capital: Vientiane. Population: 5.6 million. Currency: kip. Language: Lao.

- **Cambodia:** Capital: Phnom Penh. Population: 12.5 million. Currency: riel. Language: Khmer.

- **Indonesia:** Capital: Jakarta. Population: 212.6 million. Currency: rupiah. Main language: Bahasa Indonesia.

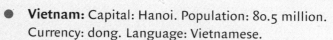

◀ Warm and damp, Southeast Asia is a fertile region where Buddhist and Hindu kings once built giant temples in the forests, but many people today are desperately poor.

- **Laos, Vietnam and Cambodia** were once French colonies and the end of French rule in the 1950s led to years of suffering and war.

- **Both Laos and Vietnam** are one-party communist states, although their governments are elected by popular vote. In Cambodia, the king was reinstated in 1993.

- **Many people** in Laos and Vietnam are poor and live by growing rice. Laos is the world's poorest country.

- **In Indonesia**, an elected president has replaced dictator General Suharto but the military still have great power.

- **Spread over 13,700 islands,** Indonesia is one of the world's most densely populated countries. Jakarta is home to 12.4 million, and is heavily industrialized. In the country hillside terraces ensure every inch is used for rice.

- **Indonesian rainforest** is being rapidly destroyed by loggers. In 1997, parts of the country were engulfed by smoke from fires started by loggers.

Malaysia and Singapore

- **Malaysia:** Capitals: Kuala Lumpur and Putrajaya. Pop: 22.3 million. Curr: Malaysian dollar (ringgit). Language: Bahasa Malaysia.
- **Singapore:** Capital: Singapore. Population: 4 million. Currency: Singapore dollar. Languages: English, Mandarin, Malay and Tamil.
- **Malaysia** is split into sections: peninsular Malaysia and Sarawak and Sabah on the island of Borneo.
- **In the 1980s** Malaysia was a farming country relying on rubber for exports.
- **Malaysia** is still the world's top rubber producer.
- **Cheap, skilled labour** and oil have turned Malaysia into one of the world's most rapidly developing economies.
- **A plan called 2020 Vision** aims to have Malaysia fully developed by the year 2020.
- **Singapore** may be the world's busiest port. Huge

▲ *Singapore is one of the busiest and most prosperous cities in Asia.*

ships tie up here every 3 minutes.

- **Singapore** is also one of Asia's most successful trading and manufacturing centres.
- **Singapore** has a state-of-the-art transport system, kept immaculately clean by strict laws governing litter.

Peoples of Europe

◀ *In East Europe, many people, like these Romanians, have their own traditional dress.*

- **About 730 million** people live in Europe – about 12% of the world's population.
- **Europe** is one of the most densely populated continents averaging 70 people per square kilometre.
- **Most Europeans** are descended from tribes who migrated into Europe more than 1500 years ago.

- **Most British people** are descended from a mix of Celts, Angles, Saxons, Danes and others. Most French people are descended from Gauls and Franks. Most Eastern Europeans are Slavic (see peoples of Northern Asia).
- **North Europeans** such as Scandinavians often have fair skin and blonde hair. South Europeans such as Italians often have olive skin and dark hair.
- **Most European countries** have a mix of people from all parts of the world, including former European colonies in Africa and Asia.
- **Most Europeans** are Christians.
- **Most Europeans** speak an Indo-European language, such as English, French or Russian.
- **Languages** like French, Spanish and Italian are romance languages that come from Latin, language of the Romans.
- **Basque people** in Spain speak a language related to no other language. Hungarians, Finns and Estonians speak a Uralic-Altaic language like those of Turkey and Mongolia.

Chile

- **Capitals:** Santiago and Valparaiso. **Area:** 756,626 sq km. **Currency:** Chilean peso. **Language:** Spanish.

- **Physical features:** Highest mountain: Ojos del Solado (6895 m). Longest river: Bio-Bio (380 km).

- **Population:** 15.2 million. Population density: 20/sq km. Life expectancy: men 72.4 years; women 78.4 years.

- **Wealth:** GDP: $95 billion. GDP per head: $6240.

- **Exports:** Copper, manufactured goods, fresh fruit.

- **Chile** is one of the world's most volcanically active countries, with 75 volcanoes. Chile also has eight of the world's highest active volcanoes, including Guallatiri and San Pedro which are both active.

- **Chile is a major** wine producer.

- **The copper mine** at Chuquicamata is the world's biggest man-made hole, 3 km wide and 750 m deep. The El Teniente copper mine is the world's deepest.

- **Chile is the** world's largest copper producer.

- **The Mapuche Indians** live in the forest area around Temuco in southern Chile and those who preserve their traditional way of life live in round straw houses.

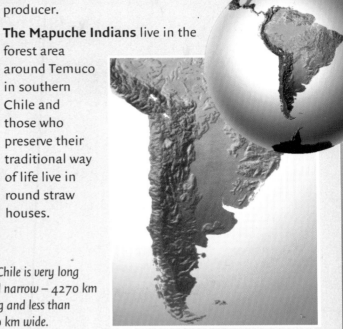

▶ Chile is very long and narrow – 4270 km long and less than 180 km wide.

Spain and Portugal

- **Spain:** Capital: Madrid. Area: 505,990 sq km. Currency: euro. Language: Spanish.

- **Physical features:** Highest mountain: Pico de Tiede (3718 m). Longest river: Tagus (1007 km).

- **Population:** 39.5 million. Population density: 71/sq km. Life expectancy: men 74.7 years; women 81.6 years.

- **Wealth:** GDP: $592.5 billion. GDP per head: $14,990.

- **Exports:** Cars, machinery, wine, fruit, steel, textiles, chemicals.

- **Much of the centre of Spain** is too hot and dry

◀ Spain and Portugal are isolated from the rest of Europe by the Pyrenees on their own peninsula, called Iberia. Mainland Europe's most westerly point, Cabo da Roca, is in Portugal.

for some crops, but perfect for olives, sunflowers and for grapes, oranges and other fruit. Spain is one of the world's leading fruit-growers.

- **Spain is one of the leading** carmakers in Europe, with huge plants in Valencia and Saragossa. It also makes a lot of iron and steel. Toledo in the south was once famous for its fine sword steel.

- **Portugal:** Capital: Lisbon. Population: 9.8 million. Currency: euro. Language: Portuguese.

- **Portugal** once had a large empire including large parts of Latin America and Africa. Yet it is fairly underdeveloped. Most people still live in the countryside, growing wheat, rice, almonds, olives and maize. Portugal is famous for its 'port', a drink made by adding brandy to wine.

> ★ STAR FACT ★
> Every summer, 63 million sunseekers come to Spain's cities, beaches and islands.

Scandinavia

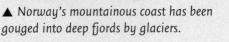
▲ Norway's mountainous coast has been gouged into deep fjords by glaciers.

- **Norway:** Capital: Oslo. Population: 4.5 million. Currency: Norwegian krone. Language: Norwegian.
- **Sweden:** Capital: Stockholm. Population: 8.9 million. Currency: Swedish krona. Language: Swedish.

- **Denmark:** Capital: Copenhagen. Population: 5.3 million. Currency: Danish krone. Language: Danish.
- **Finland:** Capital: Helsinki. Population: 5.2 million. Currency: euro. Languages: Finnish and Swedish.
- **Scandinavia** has among the iciest, most northerly inhabited countries in the world. Yet they enjoy a high standard of living and welfare provision.
  - **Norway's fishing boats** land 2.4 million tonnes of fish a year – more than those of any other European country except Russia.
  - **Sweden is known** for its high-quality engineering, including its cars such as Volvos and aircraft-makers such as Saab.
  - **Finland and Sweden** are known for their glass and ceramic work.
- **Sweden's capital Stockholm** is built on 14 islands in an archipelago comprising thousands of islands.
- **Danish farms** are famous for butter and bacon.

Political systems

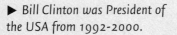
▶ Bill Clinton was President of the USA from 1992-2000.

- **Democracies** are countries with governments elected every few years by popular vote.
- **Most democracies** have a constitution, a written set of laws saying how a government must be run.
- **Democracies** like France are republics. This means the head of state is an elected president. In some republics like the USA, the president is in charge; in others, the president is a figurehead and the country is run by a chancellor or prime minister.

- **Monarchies** are countries which still have a monarch – a king or queen – like Britain. But their power is usually limited and the country is run by an elected government.
- **In autocracies** a single person or small group of people hold all the power, as in China and North Korea.
- **Most governments** are split into the legislature who make or amend laws, the executive who put them into effect and the judiciary who see they are applied fairly.
- **Most countries** are capitalist, which means most things – capital – are owned by individuals or small groups.
- **A few countries** like Cuba are communist, which means everything is owned by the community, or rather the state.
- **Socialists** believe the government should ensure everyone has equal rights, a fair share of money, and good health, education and housing.
- **Fascists** believe in rigid discipline and that they and their country are superior to others, like Hitler's Germany in the 1930s. There is no openly fascist country at present.

The Pacific Islands

- **Scattered** around the Pacific are countless islands – approximately 25,000. Some are little more than rocks; some are thousands of square kilometres.

- **The Pacific Islands** are in three main groups: Melanesia, Micronesia and Polynesia.

- **Melanesia** includes New Guinea, the Solomons, New

Caledonia, Vanuatu and Fiji.

- **Melanesia** means 'black islands' and gets its name from the dark skin of many of the islanders here.

- **Micronesia** is 2,000 islands to the north of Melanesia, including Guam and the Marshall Islands.

- **Micronesia** means 'tiny islands'.

- **Polynesia** is a vast group of islands 8000 km across. It includes Tahiti, Samoa, Tonga, Kiribati and Easter Island.

- **Polynesia** means 'many islands'.

- **Some of the islands** in the Pacific are either extinct volcanoes, or coral islands built around a volcanic peak. Atolls are coral rings left as the volcano sinks.

- **Most Pacific islanders** live in small farming or fishing villages as they have for thousands of years, but western influences are changing the island way of life rapidly.

◀ Like many Pacific islands, Fiji seems like a paradise.

Mexico

- **Capital:** Mexico City. Area: 1,958,201 sq km. Currency: Mexican peso. Language: Spanish.

- **Physical features:** Highest mountain: Volcán Citlaltépetl (5610 m). Longest river: Rio Bravo (3035 km).

- **Population:** 97 mill. Population density: 50/sq km. Life expectancy: men 69.7 years; women 75.7 years.

- **Wealth:** GDP: $469.5 billion. GDP per head: $4840.

- **Exports:** Petroleum, vehicles, machinery, cotton, coffee, fertilizers, minerals.

- **Mexico** is quite mountainous and only 13% of the land is suitable for farming, but the soil that develops on lava poured out by Mexico's many volcanoes is very fertile. Where there is enough rain, there are big plantations for tobacco, coffee, cane, cocoa, cotton and rubber.

- **Over half Mexico's** export earnings come from manufactured goods – notably cars.

- **Mexico has** a rapidly growing population. The birth rate is high and half the population is under 25.

- **Most of Mexico's people** are *mestizos* – descended from both American Indians and Europeans. But there are still 29 million American Indians.

- **Mexico City** is one of the world's biggest, busiest, dirtiest cities. The urban area has a population of over 18.4 million – and it is growing rapidly as more people move in from the country to find jobs.

◀ Mexico lies immediately south of the USA, between the Gulf of Mexico and the Pacific.

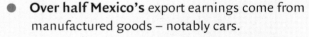

Gulf of California

Pacific Ocean

Gulf of Mexico

Yucatán Peninsular

Mexico City

Gulf of Tehuantapec

Mediterranean food

- **Mediterranean food** depends on ingredients grown in the warm Mediterranean climate. It tends to be lighter than north European food, including salads, flat bread and fish rather than sauces and stews.

- **Olive oil** is used for dressing salads and frying food.
- **Major styles** of Mediterranean food include Italian, Greek, Turkish, Spanish and North African.

◄ *Spaghetti Bolognese – spaghetti pasta with meat and tomato sauce – is the centrepiece of a typical Italian meal.*

- **Italian meals** often include pasta, which is made from durum wheat flour and served with a sauce.
- **Popular forms of pasta** include spaghetti ('little strings'), vermicelli ('little worms'), fusilli ('spindles') and tube-shaped macaroni.
- **In north Italy** ribbon pastas served with cream sauces are popular. In the south, macaroni served with tomato-based sauces are more popular.
- **Pizzas** are popular snacks, especially in the south.
- **Greek food** includes meats – especially lamb – and fish cooked in olive oil.
- **Greek salad** includes olives, cucumber, tomatoes, herbs and feta cheese (soft goat's cheese).
- **Spanish food** often includes seafood such as *calamares* (squid). *Paella* includes seafoods and chicken combined with rice and cooked in saffron. *Gazpacho* is a cold tomato soup. *Tapas* are small snacks, originating in southern Spain.

Pakistan and Bangladesh

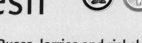

- **Pakistan:** Capital: Islamabad. Population: 156 million. Currency: Pakistan rupee. Language: Urdu.
- **Bangladesh:** Capital: Dhaka. Population: 128.3 million. Currency: taka. Language: Bengali.
- **The Punjab** region is where many Pakistanis live. It gets its name – which means 'five waters' – from five tributaries of the River Indus: the Jhelum, Chenab, Ravi, Sutlej and Beus. These rivers water the Punjab's plains and make it fertile. All the same, large areas of the Punjab are dry and rely on one of the world's biggest irrigation networks.
- **Pakistan's major exports** include textiles, cement, leather and machine tools.

► *Pakistan and Bangladesh were once one nation, East and West Pakistan, but East Pakistan broke away in 1971 to become Bangladesh.*

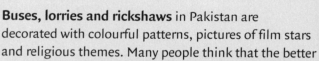

- **Buses, lorries and rickshaws** in Pakistan are decorated with colourful patterns, pictures of film stars and religious themes. Many people think that the better the vehicle looks, the more careful the driver will be.
- **Pakistan's capital** is the brand new city of Islamabad, built in the 1960s, but its biggest city and industrial centre is the port of Karachi.
- **While people in India** are mainly Hindu, in Pakistan and Bangladesh they are mainly Muslim.
- **Jute is a reed** that thrives in Bangladesh's warm, moist climate. It is used for making rope, sacking and carpet backing.
- **Over 70 big jute mills** make jute Bangladesh's most important export.
- **Most of Bangladesh** is low-lying and prone to flooding. Floods have devastated Bangladesh several times in the past 40 years.

Switzerland and Austria

- **Switzerland:** Capital: Berne. Population: 7.4 million. Currency: Franc. Languages: German, French, Italian.
- **Austria:** Capital: Vienna. Population: 8.1 million. Currency: euro. Language: German.
- **Switzerland and Austria** are small but beautiful countries mostly in the Alps mountains.
- **Both Switzerland and Austria** make a great deal of money from tourists who come to walk and ski here.
- **Switzerland** has long been 'neutral', staying out of all the major wars. This is why organizations like the Red Cross and World Health Organization are based there.
- **Switzerland is** one of the richest countries in terms of GDP per person ($38,680). Luxembourg's is $45,320.
- **People from** all over the world put their money in Swiss banks because the country is stable politically and its banking laws guarantee secrecy.
- **Switzerland is famous** for making small, valuable things such as precision instruments and watches.

- **Vienna** was once the heart of the great Austrian Empire and the music capital of Europe.
- **Austrians** rely on mountain-river hydroelectricity for much of its power.

◄ *Austria earns more than a sixth of its income from tourists who come to enjoy the Alpine scenery.*

New England

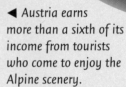

▲ *New England is famous for the stunning colours of its trees in 'fall' (autumn), when the leaves turn reds, golds and ambers.*

- **New England** is six states in northeast USA – Maine, Vermont, New Hampshire, Massachusetts, Rhode Island and Connecticut.
- **New England** was one of the first areas of North America settled by Europeans in the 1600s.

- **The USA's** oldest buildings are in New England.
- **New England** is famous for its attractive small towns with pretty 18th- and 19th-century white clapperboard houses and elegantly spired churches.
- **Vermont's** name means 'green mountain' and it has fewer urban inhabitants than any other state.
- **Basketball** was invented in Massachusetts in 1891.
- **Route 128** in Massachusetts is famed for its cutting-edge electronic technology factories.
- **Boston** is one of the USA's oldest, most cultured cities. It also has a large number of educational and research institutes. Harvard University is at Cambridge nearby. Yale is in Connecticut.
- **New Hampshire** is famous for its scenery.

★ **STAR FACT** ★
Rhode Island is the smallest state in the USA, which is why it is often called 'Little Rhody'.

1000 THINGS YOU SHOULD KNOW ABOUT

PLANTS

KEY

 How plants work

 Flowers

 Biomes

 Mosses etc.

 Trees

 Plants and humans

Perennial flowers

▲ *Chrysanthemums are among the most popular perennials.*

- **Garden perennials** are flowers that live for at least three years.

- **Perennials** may not bloom in the first year, but after that they bloom every year.

- **Since they bloom** for many years, perennials do not need to produce as many seeds to survive.

- **Some perennials** are herbaceous – that is, they have soft stems. The stems wither at the end of each summer and new stems grow next spring.

- **Woody perennials** have woody stems. Their stems don't wither, but most shed their leaves in autumn.

- **Perennials** from temperate (cool) regions, like asters, irises, lupins, wallflowers, peonies and primroses, need a cold winter to encourage new buds to grow in spring.

- **Tropical perennials** such as African violets, begonias and gloxinias cannot survive winters outdoors in temperate climates.

- **Most perennials** spread by sending out shoots from their roots which develop into new stems.

- **Some perennials** such as columbines and delphiniums last for only three or four years.

- **Gardeners** spread perennials by taking cuttings – that is, pieces cut from stems or roots.

Oak trees

- **Oaks** are a group of over 450 different trees. Most belong to a family with the Latin name *Quercus*.

- ***Quercus* oaks** grow in the northern half of the world in temperate regions or high up in the tropics.

- **Southern oaks,** such as the Australian and Tasmanian oaks, don't belong to the *Quercus* family.

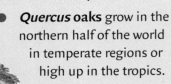

◄ *Oaks have leaves with four or five pairs of lobes. They grow fruits called acorns in a little cup.*

> ★ **STAR FACT** ★
> The bark of cork oaks in Portugal and Spain is made into corks for bottles.

- **Most oaks** from warmer places, such as the holm oak, are evergreen.

- **When a nail** is driven into freshly cut oak, it creates a blue stain as tannin in the wood reacts with the iron.

- **Tannin from oak bark** has been used for curing leather since the days of Ancient Greece.

- **Oak trees** can live a thousand years or more and grow up to 40 m. In Europe, oaks are the oldest of all trees.

- **Oak wood** is very strong and durable and so was the main building wood for centuries – used for timber frames in houses and for building ships.

- **Oak trees** are divided into white oaks like the English oak and red oaks like the North American pine oak according to the colour of their wood.

Spices

- **The Phoenicians** traded in spices 2500 years ago.

- **The great voyages** of exploration of the 1400s, like those of Columbus, were mainly to find ways to reach sources of spices in southeast Asia.

- **The Molucca Islands** in Indonesia were known as the Spice Islands because they were the main source of cloves, nutmeg and mace.

- **Sesame** was used by the Ancient Chinese for ink and by the Romans as sandwich spread. Arabs thought it had magical powers. In *Ali Baba and the 40 Thieves*, Ali says, 'open sesame' to magically open a door.

- **Cinnamon** is the inner bark of a laurel tree native to Sri Lanka. It was once more valuable than gold.

▲ *Spices made from fragrant tropical plants have long been used to flavour food.*

- **Allspice** is the berries of a myrtle tree native to the West Indies. It gets its name because it tastes like a mixture of cloves, cinnamon and nutmeg.

- **In Ancient Greece and Rome** people often paid their taxes in peppercorns.

- **Cloves are the dried buds** of a large evergreen tree that grows in the Moluccas.

- **From 200 BC** Chinese courtiers sucked cloves to make their breath smell sweet for the Emperor.

- **Saffron** is the yellow stigmas of the purple saffron crocus, used as a dye by Buddhist priests. It is the most costly of all spices. It takes 170,000 flowers to make just 1 kg.

Marine plants

     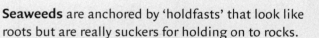

- **Plants in the sea** can only live in the sunlit surface waters of the ocean, called the photic zone.

- **The photic zone** goes down about 100 m.

- **Phytoplankton** are minute, floating, plant-like organisms made from just a single cell.

- **Almost any marine plant** big enough to be seen with the naked eye is called seaweed.

- **Seaweeds** are anchored by 'holdfasts' that look like roots but are really suckers for holding on to rocks.

- **Seaweeds** are red, green or brown algae. Red algae are small and fern-like and grow 30–60 m down in tropical seas. Brown algae like giant kelp are big and grow down to about 20 m, mostly in cold water.

- **Some seaweeds** such as the bladderwrack have gas pockets to help their fronds (leaves) float.

- **The fastest growing** plant in the sea is the giant kelp, which can grow 1 m in a single day. Giant kelp can grow up to 60 m long.

- **The Sargasso Sea** is a vast area of sea covering 5.2 million sq km east of the West Indies. Gulfweed floats so densely here that it looks like green meadows.

- **The Sargasso Sea** was discovered by Christopher Columbus in 1492.

◄ *Seaweeds don't have roots, stems, leaves or flowers, but they are plants and make their food from sunlight (see photosynthesis).*

Gardens

- **The Ancient Chinese and Greeks** grew fruit trees, vegetables and herbs in gardens for food and for medicines.

- **In the 1500s** there were five famous botanical gardens in Europe designed to study and grow herbs for medicine.

- **The first botanical gardens** were at Pisa (1543) and Padua (1545) in Italy.

▲ *Gardening has become one of the most popular of all pastimes.*

- **Carolus Clusius** set up a famous flower garden in Leiden in Holland in the late 1500s. Here the first tulips from China were grown and the Dutch bulb industry began.

- **The most famous gardener** of the 17th century was John Evelyn who set up a beautiful garden at Sayes Court in Deptford near London.

- **The Royal Botanic Gardens** at Kew near London were made famous by Sir Joseph Banks in the late 1700s for their collections of plants from around the world.

- **Today Kew Gardens** has 33,400 classes of living plants and a herbarium of dried plants with 7 million species – that's 98% of the world's plants.

- **Plants** such as rubber plants, pineapples, bananas, tea and coffee were spread around the world from Kew.

- **Lancelot 'Capability' Brown** (1716-83) was a famous English landscape gardener. He got his nickname by telling clients their gardens had excellent 'capabilities'.

- **Ornamental gardens** are ordinary flower gardens.

Cones

> ★ **STAR FACT** ★
> The largest cones are those of the sugar pine which can grow over 65 cm long.

- **Cones** are the tough little clusters of scales that coniferous trees carry their seeds in.

- **The scales** on a cone are called bracts. The seeds are held between the bracts. Bracts are thin and papery in spruces and thick and woody in silver firs.

- **Pine cone bracts** have a lump called an umbo.

- **All cones** are green and quite soft when they first form, then turn brown and hard as they ripen.

- **Cones stand upright** on the branch until they are ripe and ready to shed their seeds.

- **Most cones** turn over when ripe to hang downwards so that the seeds fall out.

- **The cones of cedars** and silver firs stay upright and the bracts drop away to release the seeds.

- **Long, hanging cones** like those of the pine and spruce hang throughout winter then release seeds in spring.

- **The monkey puzzle** tree has a unique, pineapple-shaped cone with golden spines and edible seeds.

▼ *These Scots pine cones are brown and were fertilized about three years ago. Younger cones further out on branches would have been fertilized last spring and would still be green.*

Lichen

- **Lichens** are a remarkable partnership between algae and fungi.
- **The algae** in lichen are tiny green balls which make the food from sunlight to feed the fungi.
- **The fungi make a protective** layer around the algae and hold water.
- **There are 20,000** species of lichen. Some grow on soil, but most grow on rocks or tree bark.
- **Fruticose lichens** are shrub-like, foliose lichens look like leaves, and crustose lichens look like crusts.
- **Lichens only grow** when moistened by rain.
- **Lichens can survive** in many places where other plants would die, such as the Arctic, in deserts and on mountain tops.
- **Some Arctic lichens** are over 4000 years old.
- **Lichens are very sensitive** to air pollution, especially sulphur dioxide, and are used by scientists to indicate air pollution.

▲ Lichens are tiny and slow-growing – some growing only a fraction of a millimetre a year. But they are usually long-lived.

- **The oakmoss lichen** from Europe and North Africa is added to most perfumes and after-shaves to stop flower scents fading. Scandinavian reindeer moss is a lichen eaten by reindeer. It is exported to Germany for decorations.

Spores and seeds

▶ New sycamore trees grow from their tiny winged seeds (top). Mushrooms (below right) grow from spores.

- **Seed plants** are plants that grow from seeds.
- **Seeds** have a tiny baby plant inside called an embryo from which the plant grows plus a supply of stored food and a protective coating.
- **Spores contain** special cells which grow into new organisms. Green plants like ferns and mosses, and fungi like mushrooms, produce spores.
- **All 250,000 flowering plants** produce 'enclosed' seeds. These are seeds that grow inside sacs called ovaries which turn into a fruit around the seed.

- **The 800 or so** conifers, cycads and gingkos produce 'naked' seeds, which means there is no fruit around them.
- **Seeds** only develop when a plant is fertilized by pollen.
- **The largest seeds** are those of the double coconut or coco-de-mer of the Seychelles which can sometimes weigh up to 20 kg.
- **30,000 orchid seeds** weigh barely 1 gm.
- **The world's biggest tree,** the giant redwood, grows from tiny seeds that are less than 2 mm long.
- **Coconut trees** produce only a few big seeds; orchids produce millions, but only a few grow into plants.

Parts of a tree

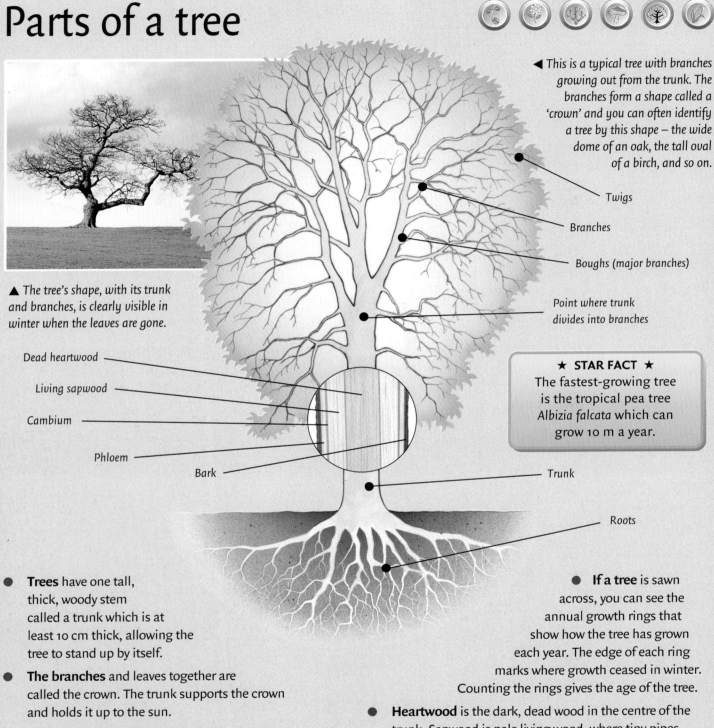

▲ The tree's shape, with its trunk and branches, is clearly visible in winter when the leaves are gone.

◀ This is a typical tree with branches growing out from the trunk. The branches form a shape called a 'crown' and you can often identify a tree by this shape – the wide dome of an oak, the tall oval of a birch, and so on.

Twigs

Branches

Boughs (major branches)

Point where trunk divides into branches

Dead heartwood

Living sapwood

Cambium

Phloem

Bark

Trunk

Roots

> ★ **STAR FACT** ★
> The fastest-growing tree is the tropical pea tree *Albizia falcata* which can grow 10 m a year.

- **Trees** have one tall, thick, woody stem called a trunk which is at least 10 cm thick, allowing the tree to stand up by itself.

- **The branches** and leaves together are called the crown. The trunk supports the crown and holds it up to the sun.

- **The trunks of conifers** typically grow right to the top of the tree. The lower branches are longer because they have been growing longest. The upper branches are short because they are new. So the tree has a conical shape.

- **Trees with wide flat leaves** are called broad-leaved trees. They usually have crowns with a rounded shape.

- **The trunk and branches** have five layers from the centre out: heartwood, sapwood, cambium, phloem and bark.

- **If a tree** is sawn across, you can see the annual growth rings that show how the tree has grown each year. The edge of each ring marks where growth ceased in winter. Counting the rings gives the age of the tree.

- **Heartwood** is the dark, dead wood in the centre of the trunk. Sapwood is pale living wood, where tiny pipes called xylem carry sap from the roots to the leaves.

- **The cambium** is the thin layer where the sapwood is actually growing; the phloem is the thin food-conducting layer.

- **The bark** is the tree's protective skin of hard dead tissue. Bark takes many different forms and often cracks as the tree grows, but it is always made from cork.

Roses

- **The rose** is one of the most popular of all garden flowers because of its lovely perfume and beautiful blooms.
- **Wild roses** usually have small flowers and have a single layer of five petals. Garden roses usually have big flowers with multiple sets of five petals in two or more layers.
- **There are 100 species** of wild rose, but all today's garden roses were created by crossing 10 Asian species.
- **There are now over 13,000** official varieties of garden rose altogether.
- **Some experts divide garden roses** into groups by when they bloom: old roses bloom once a year in early summer; perpetual roses bloom in early summer, then again in autumn; and everblooming hybrids bloom all summer.
- **Old roses** include yellow briers, damask roses and many climbing roses.
- **Perpetuals** include what are called hybrid perpetuals.
- **Everblooming hybrids** include floribundas, hybrid teas,

▶ Roses often look their best just after they begin to open, when the petals are still in a tight, velvety cluster.

gloribundas and polyanthas.

- **Hybrid teas** such as the Peace are the most popular of all roses. They were created by crossing everblooming but fragile tea roses with vigorous hybrid perpetuals.
- **Attar of roses** is a perfume made from roses, especially damask roses.

Rice

- **Rice** is a cereal grain that is the basic food of half the world's population. It is especially important in SE Asia.
- **The wild rice** or Indian rice collected by North American Indians for thousands of years is not related to rice.
- **Like other cereals,** rice is a grass, but it grows best in shallow water in tropical areas.
- **Rice growers** usually flood their fields to keep them wet. The flooded fields are called paddies.

> ★ STAR FACT ★
> A lot of wheat is fed to livestock, but 95% of all rice is eaten by people.

- **The rice seeds** are sown in soil, then when the seedlings are 25–50 days old they are transplanted to the paddy field under 5–10 cm of water.
- **Brown rice** is rice grain with the husk ground away. White rice is rice grain with the inner bran layer ground away as well, and is far less nutritious.
- **Rice-growing** probably began in India about 3000BC.
- **In 1962** researchers in the Philippines experimented with hybrids of 10,000 strains of rice. They made a rice called 'IR-8' by crossing a tall, vigorous rice from Indonesia and a dwarf rice from Taiwan.
- **IR-8** sometimes gave double yields, and was called 'miracle rice', but it did not grow well in poor soils .

◀ To keep paddies flooded, fields on hillsides are banked in terraces.

Mushrooms

▲ Like other fungi, mushrooms cannot make their own food and feed off hosts such as trees.

- **Mushrooms** are umbrella-shaped fungi, many of which are edible.
- **Mushrooms** feed off either living or decaying plants.
- **Poisonous mushrooms** are called toadstools.
- **The umbrella-shaped** part of the mushroom is called the fruiting body. Under the surface is a mass of fine stalk threads called the mycelium.
- **The threads** making up the mycelium are called hyphae (said hi-fi). These absorb food.
- **The fruiting body** grows overnight after rain and lasts just a few days. The mycelium may survive underground for many years.
- **The fruiting body** is covered by a protective cap. On the underside of the cap are lots of thin sheets called gills which are covered in spores.
- **A mushroom's** gills can produce 16 billion spores in its brief lifetime.
- **The biggest mushrooms** have caps up to 50 cm across and grow up to 40 cm tall.
- **Fairy rings** are rings of bright green grass once said to have been made by fairies dancing. They are actually made by a mushroom as its hyphae spread outwards. Chemicals they release make grass grow greener. Gradually the mycelium at the centre dies while outer edges grow and the ring gets bigger.

Forestry

▲ The signs of pollarding are easy to see in these trees in winter when the leaves are gone.

- **Forests** provide fuel, timber, paper, resins, varnishes, dyes, rubber, kapok and much more besides.
- **Softwood** is timber from coniferous trees such as pine, larch, fir and spruce. 75–80% of the natural forests of northern Asia, Europe and the USA are softwood.

> ★ STAR FACT ★
> Every year the world uses three billion cubic metres of wood – a pile as big as a football stadium and as high as Mt Everest.

- **In vast plantations** fast-growing conifers are set in straight rows so they are easy to cut down.
- **Hardwood** is timber from broad-leaved trees such as oak. Most hardwood forests are in the tropics.
- **Hardwood trees** take over a century to reach maturity.
- **Tropical hardwoods** such as mahogany are becoming rare as more hardwood is cut for timber.
- **Pollarding** is cutting the topmost branches of a tree so new shoots grow from the trunk to the same length.
- **Coppicing** is cutting tree stems at ground level to encourage several stems to grow from the same root.
- **Half the world's remaining** rainforests will be gone by 2020 if they are cut down at today's rate.

Sugar

- **Sugars** are sweet-tasting natural substances made by plants and animals. All green plants make sugar.

- **Fruit and honey** contain a sugar called fructose. Milk contains the sugar lactose.

▼▶ *Crystals of demerara sugar are made from the sugary juice from the stems of the tropical sugar cane.*

- **The most common sugar** is called sucrose, or just sugar – like the sugar you sprinkle on cereal.

- **Sugar is made** from sugar cane and sugar beet.

  - **Sugar cane** is a tropical grass with woody stems 2–5 m tall. It grows in places like India and Brazil.

    - **Sugar juice is made** from cane by shredding and crushing the stems and soaking them in hot water to dissolve the sugar.

    - **Sugar beet** is a turnip-like plant that grows in temperate countries.

    - **Sugar juice is made** from beet by soaking thin slices of the root in hot water to dissolve the sugar.

- **Sugar juice** is warmed to evaporate water so crystals form.

- **White sugar** is sugar made from sugar beet, or by refining (purifying) cane-sugar. Brown sugars such as muscovado and demerara are unrefined cane-sugar. Molasses and black treacle are by-products of cane-sugar refining.

Leaves

- **Leaves** are a plant's powerhouse, using sunlight to join water and carbon dioxide to make sugar, the plant's fuel.

- **Leaves are** broad and flat to catch maximum sunlight.

- **Leaves** are joined to the stem by a stalk called a petiole.

- **The flat part** of the leaf is called the blade.

- **The leaf blade** is like a sandwich with two layers of cells holding a thick filling of green cells.

- **The green** comes from the chemical chlorophyll. It is this that catches sunlight to make sugar in photosynthesis.

- **Chlorophyll** is held in tiny bags in each cell called chloroplasts.

- **A network** of branching veins (tubes) supplies the leaf with water. It also transports the sugar made there to the rest of the plant.

- **Air containing** carbon dioxide is drawn into the leaf through pores on the underside called stomata. Stomata also let out water in a process called transpiration.

- **To cut down water loss** in dry places, leaves may be rolled-up, long and needle-like, or covered in hairs or wax. Climbing plants, such as peas, have leaf tips that coil into stalks called tendrils to help the plant cling.

▶ *A hugely magnified slice through a leaf, showing the cells and veins.*

Leaf veins containing tiny tubes

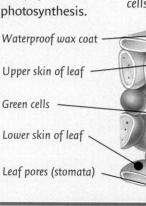
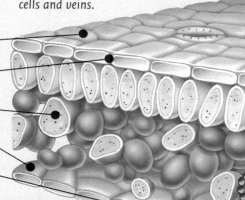

Waterproof wax coat

Upper skin of leaf

Green cells

Lower skin of leaf

Leaf pores (stomata)

Rotting trees

- **Trees** are dying in forests all the time.

- **In the past** foresters used to clear away dead trees or chop down those that were dying, but it is now clear that they play a vital part in the woodland ecosystem.

- **When a tree falls** it crashes down through the leaves and opens up a patch of woodland, called a glade, to the sky.

- **In the glade** saplings (new young trees) can sprout and flourish in the sunlight.

- **Many other woodland plants** flourish in the sunshine of a glade.

- **Flowers** such as foxgloves and rosebay willowherbs often spring up in a glade.

- **Bracken and shrubs** such as brambles grow quickly in a glade.

- **The rotting tree trunk** provides food for fungi such as green-staining and candle snuff fungus.

- **Many insects** such as beetles find a home in the rotting wood.

- **As the rotting tree is broken down** it not only provides food for plants, insects and bacteria; it enriches the soil too.

▶ Rotting trees provide a home for many kinds of plants, such as these tiny green liverworts, growing on an old stump.

Tulips

▲ Huge numbers of tulips are now grown in the fields in Holland.

- **Tulips** are flowers that bloom in spring from bulbs.

- **Tulips are** monocots and produce one large, bell-shaped bloom at the end of each stem.

- **There are about** 100 species of wild tulip, growing right across Asia to China.

- **Tulips** come in most colours but blue. Reds and yellows are common, but they vary from white to deep purple.

- **There are over 4000** garden varieties.

- **Most tulips** are 'late bloomers' with names like breeders, cottages and parrots.

- **Mid-season bloomers** include Mendels and Darwins.

- **Early season** bloomers include single-flowereds and double-flowered earlies.

- **Tulips** were introduced to Europe in 1551 by the Viennese ambassador to Turkey, Augier de Busbecq. But Holland became the centre of tulip-growing early in the 1600s, when Europe was gripped by 'tulipmania'. At this time, people would exchange mansions for a single tulip bulb. Holland is still the centre of tulip growing.

> ★ STAR FACT ★
> The word *tulip* comes from the Turkish for 'turban', because of their shape.

Herbs

- **Herbs** are small plants used as medicines or to flavour food.

- **Most herbs** are perennial and have soft stems which die back in winter.

- **With some herbs** such as rosemary, only the leaves are used. With others, such as garlic, the bulb is used. Fennel is used for its seeds as well as its bulb and leaves. Coriander is used for its leaves and seeds.

- **Basil** gets its name from the Greek *basilikon* or 'kingly', because it was so highly valued around the Mediterranean for its strong flavour. In the Middle Ages, judges and officials used to carry it in posies to ward off unpleasant smells.

- **Rosemary** is a coastal plant and gets its name from the Latin *ros marinus*, meaning 'sea dew'. People who study herbs – herbalists – once thought it improved memory.

- **Bay leaves** are the leaves of an evergreen laurel tree. They were used to make crowns for athletes, heroes and poets in Ancient Rome. It is said that a bay tree planted by your house protects it from lightning.

- **Oregano** or marjoram is a Mediterranean herb used in Italian cooking. The plant gave its name to the American state of Oregon where it is now very common.

- **Sage** is a herb thought by herbalists of old to have special healing qualities. Its scientific name *Salvia* comes from the Latin word *salvere*, 'to save'.

- **St John's wort** is a perennial herb with yellow flowers which was said to have healing qualities given by St John the Baptist. The red juice of its leaves represented his blood. Now many people use it to treat depression.

> ★ STAR FACT ★
> The root of the mandrake was supposed to have magical properties. Anyone who uprooted one was said to die, so people tied the root to a dog's tail to pull it up.

▶ These are just some of the more common herbs used in cooking, either fresh or dried. The flavour comes from what are called 'essential oils' in the leaves. Parsley, thyme and a bay leaf may be tied up in a piece of muslin cloth to make what is called a bouquet garni. This is hung in soups and stews while cooking to give them extra flavour, but is not actually eaten.

Thyme

Rosemary

Mint

Dill

Parsley

Bay

Sage

Chives

Fennel

Monocotyledons

- **Monocotyledons** are one of the two basic classes of flowering plant. The other is dicotyledons.

- **Monocotyledons** are plants that sprout a single leaf from their seeds.

- **Monocotyledons** are also known as monocots or Liliopsida.

- **There are about** 50,000 species of monocots – about a quarter of all flowering plants.

- **Monocots** include grasses, cereals, bamboos, date palms, aloes, snake plants, tulips, orchids and daffodils.

- **Monocots** tend to grow quickly and their stems stay soft and pliable, except for bamboos and palms. Most are herbaceous.

▲ Daffodils are typical monocots, with long lance-like leaves and petals in threes.

- **The tubes or veins** in monocot leaves run parallel to each other. They also develop a thick tangle of thin roots rather than a single long 'tap' root, like dicots.

- **The flower parts of monocots** such as petals tend to be set in threes or multiples of three.

- **Unlike dicots,** monocot stems grow from the inside. Dicots have a cambium, which is the layer of growing cells near the outside of the stem. Monocots rarely have a cambium.

- **Monocots** are thought to have appeared about 90 million years ago, developing from water lily-like dicots living in swamps and rivers.

Medicinal plants

▲ Aspirin is the painkiller most widely used today. It first came from the bark of willow trees.

- **Prehistoric neanderthal people** probably used plants as medicines at least 50,000 years ago.

- **Until quite recently** herbaceous plants were our main source of medicines. Plants used as medicines were listed in books called herbals.

- **An Ancient Chinese** list of 1892 herbal remedies drawn up over 3000 years ago is still used today.

- **The famous illustrated herbal** of Greek physician Dioscorides was made in the 1st century BC.

- **The most famous English** herbalist was Nicholas Culpeper, who wrote A Physical Directory in 1649.

- **Most medicines,** except antibiotics, come from flowering plants or were first found in flowering plants.

- **Powerful painkilling** drugs come from the seeds of the opium poppy.

- **Digitalis** is a heart drug that came from foxgloves. It is poisonous in large doses.

- **Garlic** is thought to protect the body against heart disease – and vampires!

> ★ STAR FACT ★
> Vincristine is a drug made from the Madagascar periwinkle that helps children fight cancer.

Pine trees

- **Pine trees** are evergreen conifers with long needle-like leaves. They grow mostly in sandy or rocky soils in cool places.

- **Pines** are the largest family of conifers.

- **There 90–100 species of pine** – most of them coming originally from northern Eurasia and North America.

- **Pines grow** fast and straight, reaching their full height in less than 20 years – which is why they provide 75% of the world's timber.

- **Some pines** produce a liquid called resin which is used to make turpentine, paint and soap.

- **Soft or white pines**, such as sugar pines and piñons, have soft wood. They grow needles in bundles of five and have little resin.

- **Hard or yellow pines**, such as Scots, Corsican and loblolly pines, have harder wood. They grow needles in bundles of two or three and make lots of resin.

◀ Like all conifers, Corsican pines produce cones. The cones look very different from flowers but serve the same purpose – making the seeds from which new trees grow.

- **Eurasian pines** include the Scots pine, Corsican pine, black pine, pinaster and stone pine.

- **North American pines** include the eastern white pine, sugar pine, stone pines, piñons, Ponderosa pine, and Monterey pine.

- **The sugar pine** is the biggest of all pines, often growing up to 70 m tall and 3.5 m thick. The eastern white pine has valuable fine white wood.

Tropical fruit

- **Tropical fruits** grow mainly in the tropics where it is warm because they cannot survive even a light frost.

- **The best-known tropical fruits** are bananas and pineapples. Others include guavas, breadfruit, lychees, melons, mangoes and papayas.

- **Banana plants** are gigantic herbs with trunks that grow 3–6 m high.

- **Alexander the Great** saw bananas in India in 326BC. Bananas were taken to the Caribbean from the Canaries c. 1550. They are now one of the main Caribbean crops.

- **There are hundreds** of varieties of banana. Most widely used is the Gros Michel. Plantains are cooking bananas.

- **Pineapples** come from Central America, and were seen by Columbus and Sir Walter Raleigh.

- **The Portuguese** took pineapples to India about 1550. Thailand is now the world's leading producer.

- **Mangoes** grow on evergreen trees of the cashew family in Burma and India.

- **The mango** is sacred to Buddhists because the mango groves provided welcome shade for Buddha.

- **Melons** are a huge group of big, round fruit with soft, juicy flesh, including canteloupes. They grow on trailing vines. Watermelons are not true melons.

▼ Bananas are picked green and unripe, shipped in refrigerated ships, then artificially ripened with 'ethylene' gas to turn them yellow.

Tropical rainforest

Tall, isolated trees called emergents grow up to 6om tall

Main canopy of broad-leaved evergreen trees

Plants called epiphytes grow on the branches of trees

Climbing lianas

Dense understorey of shrubs

- **Tropical rainforests** are warm and wet, with over 2,000 mm of rain a year and average temperatures over 20°C. This is why they are the world's richest plant habitats.

- **Flowering plants** (angiosperms) originated in tropical rainforests. Eleven of the 13 oldest families live here.

- **Most rainforest trees** are broad-leaved and evergreen.

- **Trees** of the Amazon rainforest include rosewood, Brazil nut and rubber, plus myrtle, laurel and palms. Trees in the African rainforest include mahogany, ebony, limba, wenge, agba, iroko and sapele.

> ★ **STAR FACT** ★
> One 23 hectare area of Malaysian rainforest has 375 species of tree with trunks thicker than 91 cm.

- **Many rainforest plants** have big, bright flowers to attract birds and insects in the gloom. Flowers pollinated by birds are often red, those by night-flying moths white or pink and those by day-flying insects yellow or orange.

- **The gloom** means many plants need big seeds to store enough food while they grow. So they grow fragrant fruits that attract animals to eat the fruit and spread the seed in their body waste. Fruit bats are drawn to mangoes. Orang-utans and tigers eat durians.

- **Many trees** grow flowers on their trunks to make them easy for animals to reach. This is called cauliflory.

- **Rainforest trees** are covered with epiphytes – plants whose roots never reach the soil but take water from the air.

- **Many plants are parasitic** including mistletoes and Rafflesia. They feed on other plants.

◀ *Most tropical rainforests have several layers. Towering above the main forest are isolated emergent trees up to 60 m tall. Below these, 30–50 m above the ground, is a dense canopy of leaves and branches at the top of tall, straight trees. In the gloom beneath is the understorey where young emergents, small conical trees and a huge range of shrubs grow. Clinging lianas wind their way up through the trees and epiphytes grow high on tree branches and trunks where they can reach daylight.*

The first crops

> ★ STAR FACT ★
> Beans, bottle gourds and water chestnuts were grown at Spirit Cave in Thailand 11,000 years ago.

- **The first crops** were probably root crops like turnips. Grains and green vegetables were probably first grown as crops later.

- **Einkorn and emmer** wheat and wild barley may have been cultivated by Natufians (stone-age people) about 7000BC at Ali Kosh on the Iran-Iraq border.

- **Pumpkins** and beans were cultivated in Mexico c.7000BC.

- **People** in the Amazon have grown manioc to make a flat bread called *cazabi* for thousands of years.

- **Corn** was probably first grown about 9000 years ago from the teosinte plant of the Mexican highlands.

- **Russian botanist** N. I. Vavilov worked out that wheat and rye came from the wild grasses of central Asia, millet and barley from highland China and rice from India.

- **Millet** was grown in China from c.4500BC.

- **In N. Europe** the first grains were those now called fat hen, gold of pleasure and curl-topped lady's thumb.

- **Sumerian** farmers in the Middle East c.3000BC grew barley along with wheat, flax, dates, apples, plums and grapes.

▲ Emmer wheat is one of the oldest of all cereal crops. It was probably first sown deliberately from wild grass seeds about 10,000 years ago.

Magnolia

▲ Magnolias have a single large typically white or pink flower at the end of each stem. This is the evergreen Magnolia kobus.

- **Magnolias** are evergreen shrubs, climbers and trees.

- **Magnolias** are named after the French botanist Pierre Magnol (1638-1715).

- **They produce** beautiful large white or pink flowers and are popular garden plants.

- **Nutmegs, custard** apples, ylang-ylangs and tulip trees are all kinds of magnolia.

- **There are** over 80 different kinds of magnolia.

- **Magnolias may be** the most ancient of all flowering plants. Their fossil remains have been found in rocks 120 million years old – when the dinosaurs lived.

- **A seed 2,000 years old,** found by archaeologists (people who study ancient remains) in Japan was planted in 1982. It grew and produced an unusual flower with eight petals.

- **The most popular** garden magnolia was bred in a garden near Paris, France, from a wild Japanese kind (*Magnolia liliiflora*) and a wild Chinese kind (*Magnolia denudata*).

- **Magnolia trees** have the largest leaves and flowers of any tree outside the tropical forests.

- **The cucumber tree** – a kind of magnolia – is named after its seed clusters, which look like cucumbers.

Tundra

▲ In spring the tundra bursts into glorious colour as flowers bloom to take advantage of the brief warm weather.

- **Tundras** are regions so cold and with so little rain that tall trees cannot grow.

- **Tundras** are typically covered in snow for at least half the year. Even in summer the soil 1 m or so below the ground may be permanently frozen.

- **The frozen ground** stops water draining away and makes tundras marshy and damp.

- **Winter temperatures** in the tundra can drop to -40°C. Even summer temperatures are rarely above 12°C on average.

- **Mosses and lichens**, grasses and sedges, heathers and low shrubs grow in tundra. Trees only grow in stunted forms such as dwarf willows and ash trees.

- **In spring** tundra plants grow quickly and bright wildflowers spread across the ground.

- **Arctic tundras** occur in places like northern Siberia and Canada.

- **Alpine tundras** occur high on mountains everywhere.

- **Arctic flowers** include saxifrages, Arctic poppies, fireweeds, cinquefoil, louseworts and stonecrops.

- **Alpine flowers** are often the same as Arctic flowers. They include mountain avens, gentians, saxifrages and snowbells.

Coffee and tea

- **Coffee** comes from the glossy, evergreen *Coffee arabica* shrub which originally grew wild in Ethiopia. Coffee is now grown in tropical countries around the world.

- **The coffee plant** is a mountain plant and grows best from about 1000 to 2500 m up.

- **Coffee beans** are not actually beans at all; they are the seeds inside red berries.

- **Coffee plants** can grow over 6 m tall, but they are usually pruned to under 4 m to make picking easier.

- **A coffee plant** yields only enough berries to make about 0.7 kg of coffee each year.

- **Coffee berries** are picked by hand then pulped to remove the flesh and finally roasted.

▲ Coffee berries appear green at first, then turn yellow and eventually bright red as they ripen.

- **Tea** is the leaves of the evergreen tea plant that grows in the tropics, mostly between 1000 and 2000 m.

- **Tea plants** have small, white, scented flowers and nuts that look like hazelnuts.

- **Tea plants** grow 9 m tall but they are pruned to 3 m.

> ★ STAR FACT ★
> Legend says Ethiopian goatherds discovered coffee when they saw their goats staying awake all night after eating the berries of the coffee plant.

Maple trees

- **Maples** are a huge group of trees belonging to the Acer family.

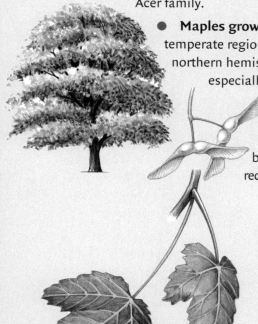

- **Maples grow** all over the temperate regions of the northern hemisphere, but especially in China.

- **Many maple tree leaves** turn brilliant shades of red in autumn.

◀ All maple trees have winged seeds called samaras or keys. Many also have three-lobed leaves, like these of the sugar maple.

- **Several North American maple trees,** including the sugar maple and the black maple, give maple syrup.

- **Maple syrup** is 'sweet-water' sap. This is different from ordinary sap and flows from wounds during times of thaws when the tree is not growing. Syrup is collected between mid January and mid April.

- **Maple syrup** was used by the Native Americans of the Great Lakes and St Lawrence River regions long before Europeans arrived in North America.

- **About 30 litres** of sap give 1 litre of maple syrup.

- **The leaf of the sugar maple tree** is Canada's national symbol.

- **Many small maples** are grown as garden plants. Japanese maples have been carefully bred over the centuries to give all kinds of varieties with different leaf shapes and colours.

- **The red maple** is planted in many North American cities for its brilliant red autumn leaves.

Wildflowers

- **All flowers** were originally wild. Garden flowers have been bred over the centuries to be very different from their wild originals.

- **Wildflowers** are flowers that have developed naturally.

- **Most wildflowers** are smaller and more delicate than their garden cousins.

- **Each** kind of place has its own special range of wildflowers, although many wildflowers have now been spread to different places by humans.

- **Heathlands** may have purple blooms of heathers, yellow gorse and scarlet pimpernel.

- **In meadow grass** flowers like buttercups, daisies, clover, forget-me-nots and ragged robin often grow.

- **In deciduous woodlands** flowers like bluebells, primroses, daffodils and celandines grow.

- **By the sea** among the rocks, sea campion and pink thrift may bloom, while up on the cliffs, there may be birdsfoot trefoil among the grasses.

- **As humans** take over larger and larger areas of the world, and as farmers use more and more weedkillers on the land, many wildflowers are becoming very rare. Some are so rare that they are protected by law.

- **The lady's slipper orchid** grows only in one secret place in Yorkshire in England.

▶ There are now very few meadows with rich displays of wildflowers like this.

Arctic plants

- **The Arctic circle** is icy cold and dark for nine months of the year, but for a few months in summer it is daylight almost all the time.

- **Over 900 species** of plants cope with the Arctic climate.

- **Full-size trees** are rare in the Arctic; but grasses and sedges, mosses and lichens are common.

- **Willow trees** grow in the Arctic, but because of the cold and fierce wind, they grow less than 10 cm tall, spreading out along the ground instead.

- **Many Arctic** plants are evergreen so they are ready to make the most of the brief summer.

- **Many small** flowers are specially

- adapted to survive Arctic conditions, such as saxifrages, avens, stonecrops, snowbells and willowherbs.

- **The Arctic poppy** is the flower that blooms nearest the North Pole.

- **Butterflies and bees** are rare in the Arctic, so many plants, like mustard, rely on the wind for pollination.

- **The soil is so poor** in the Arctic that seeds make the most of any animal corpse, such as that of a musk ox. Arctic flowers often spring up inside skulls and near bones.

- **Some plants** have dark leaves and stems to soak up the sun's warmth quickly and so melt the snow.

◀ In summer, the Arctic bursts into brief life with tiny flowers and ground berries like bilberries.

Pollination

- **For seeds** to develop, pollen from a flower's male anther must get to a female stigma.

- **Some flowers are** self-pollinating and the pollen moves from an anther to a stigma on the same plant.

- **In cross-pollinating** flowers, the pollen from the anthers must be carried to a stigma on a different plant of the same kind.

- **Some pollen** is carried by the wind.

- **Most pollen** is carried on the bodies of insects such as bees or by birds or bats that visit the flower.

- **Insect-pollinated flowers** are often brightly coloured and sweet-smelling to attract bees and butterflies.

- **Bees and butterflies** are also drawn by the flower's sweet juice or nectar. As they sip the nectar, they may brush pollen on to the stigma, or take some on their bodies from anthers to the stigma of other flowers.

- **Bees and butterflies** are drawn to blue, yellow and pink flowers. White flowers draw night-flying moths.

- **Many flowers** have honey guides – markings to guide the bees in. These are often invisible to us and can only be seen in ultraviolet light, which bees can see.

- **The cuckoopint** smells like cow-dung to attract the flies that carry its pollen.

◀ Many flowers rely on attracting bees to carry their pollen.

The farming year

A few days after the harvest the soil is cultivated to get rid of unwanted weeds

Harvest time in late summer when the wheat is harvested

After cultivation, the soil is prepared by ploughing and harrowing

About six weeks after the harvest, the seeds are sown in the prepared soil

The seeds sprout before winter sets in, but don't begin to ripen until the following spring

▶ This illustration shows the typical sequence of events in a grain farmer's year, from ploughing and sowing the seed right through to harvest the following autumn.

- **The farming year** varies considerably around the world, and farmers do different tasks at different times of year in different places.

- **In temperate regions** the crop farmer's year starts in autumn after the harvest. Once the straw has been baled and the surplus burned, the race starts to prepare the soil for next year's crops before snow and frost set in.

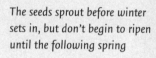

◀ Without artificial fertilizers, the soil is quickly exhausted if grain crops are planted year after year. So in the past, farmers rotated fields with different crops to allow the soil to rest. Rotation systems varied, but usually included grain, green plants, and 'rest' crops. The earliest systems had just two alternating fields. Medieval farmers used three fields. From the 1700s, rotations became more complex.

- **A tractor** drags a cultivator (like a large rake) across the field to make weeds 'chit' (germinate). A few days later the soil is cultivated again to uproot the seedlings.

- **Next the soil** is ploughed to break up the soil ready for the seed to be sown, then harrowed to smooth out the deep furrows made by the plough.

- **Within six weeks** of the harvest if the weather holds, winter wheat or barley seed is sown, fertilizer is applied and the seed soon sprouts like a carpet of grass.

- **In winter,** the farmer turns to tasks like hedge-cutting, ditching and fencing.

- **In spring** potatoes, oats and spring wheat and barley are planted, and winter crops treated with nitrogen fertilizer. In spring and summer, many farmers treat crops with 30 or more pesticides and weedkillers.

- **As the summer wears on** the wheat turns gold and is ready for harvesting when the electronic moisture metre shows it contains less than 18% moisture.

- **If the summer** is damp, the grain's moisture content may not go down below 25%, making harvesting difficult. But warm sun can quickly rectify the situation.

- **The farming year** ends with the harvest. In the past this used to be separated into various stages – harvesting, threshing, winnowing and baling. Now combine harvesters allow the farmer to complete them all at one go.

Boreal forest

- **Forests in cool regions** in the north of Asia and North America, bordering on the Arctic circle, are called boreal forests. The word *boreal* means 'northern'.

- **Winters** in boreal regions are long and cold. Days are short and snow lies permanently on the ground.

- **In Russia and Siberia** boreal forest is called *taiga*, which is Russian for 'little sticks'.

- **Boreal forests** are mostly evergreen conifers such as pine – especially Scots pine – spruce, larch and fir.

- **In Europe** boreal forests include Norway spruce and Sukaczev larch. In Siberia, there are trees such as Siberian larch and fir, chosenia and Siberian stone pine.

- **North American** forests include balsam firs, black spruces, jack pines and lodgepole pines.

> ★ STAR FACT ★
> Half the ground under conifers is covered in moss and lichen.

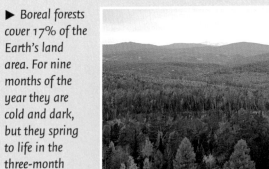

▶ *Boreal forests cover 17% of the Earth's land area. For nine months of the year they are cold and dark, but they spring to life in the three-month summer.*

- **Boreal forest floors** are covered with carpets of needles. Twinflowers, calypso orchids, lingonberries, baneberries, and coral roots are among the few plants that will grow.

- **Boreal forest trees** are good at recovering after fire. Indeed jack pine and black spruce cones only open to release their seeds after a fire.

- **The Black Dragon Fire** of 1987 in the boreal forests of China and Russia was the biggest fire in history.

The Green Revolution

- **Since ancient times** farmers have tried to improve crops. They brushed pollen from one species on to another to gain desirable qualities in the next generation of plants.

- **In 1876** Charles Darwin discovered that inbreeding – pollinating with almost identical plants – made plants less vigorous. Cross-breeding between different strains produced healthier plants.

- **In the early 1900s** American scientists found that they could improve the protein content of corn by inbreeding – but the yield was poor.

- **In 1917** Donald Jones discovered the 'double-cross', combining four strains (not the normal two) to create a hybrid corn giving high yield and high protein.

▶ *Forty years ago, many farmers abandoned traditional wheat seeds and began planting big 'superwheat' seeds.*

- **Hybrid corn** changed US farming, raising yields from 2000 litres per hectare in 1933 to 7220 in 1980.

- **In the 1960s** US farmers began growing wheat crosses such as Gaines, developed by Norman Borlaug from Japanese dwarf wheats.

- **Gaines and Nugaines** are short-stemmed wheats that grow fast and give huge yields – but they need masses of artificial fertilizers and pesticides.

- **In India and Asia** new dwarf wheats and rices created a 'Green Revolution', doubling yields in the 1960s and 1970s.

- **The Green Revolution** means farmers now use ten times as much nitrogen fertilizer as in 1960.

- **The huge cost** of special seeds, fertilizers and pesticides has often meant that only big agribusinesses can keep up, forcing small farmers out of business.

Mosses

- **Mosses** are tiny, green, non-flowering plants found throughout the world. They form cushions just a few millimetres thick on walls, rocks and logs.

- **Unlike other plants** they have no true roots. Instead, they take in moisture from the air through their stems and tiny, root-like threads called rhizoids.

- **Mosses reproduce** from minute spores in two stages.

- **First** tadpole-like male sex cells are made on bag-like stems called antheridae and swim to join the female eggs on cup-like stems called archegonia.

- **Then** a stalk called a sporophyte grows from the ova. On top is a capsule holding thousands of spores.

- **When the time** is right, the sporophyte capsule bursts, ejecting spores. If spores land in a suitable place, male and female stems grow and the process begins again.

- **Mosses** can survive for weeks without water, then soak it up like a sponge when it rains.

▶ *Mosses grow on rocks in damp places everywhere. They take in the moisture they need to grow from the air, but to reproduce, they need to be completely soaked.*

- **The sphagnum or peat moss** can soak up 25 times its own weight of water.

- **Male cells** can only swim to female cells if the moss is partly under water. So mosses often grow near streams where they get splashed.

- **Spanish moss** was often used as a filler in packing cases and to pad upholstery.

Rhododendrons

- **Rhododendrons** are a big group of 800 different trees and shrubs which belong to the heath family.

- **The word 'rhododendron'** means 'rose tree'.

- **Most rhododendrons** came originally from the Himalayas and the mountains of Malaysia where they form dense thickets.

- **Many rhododendrons** are now widely cultivated for their big red or white blooms and evergreen leaves.

- **There are over 6000** different cultivated varieties of rhododendron.

- **The spectacular June blooming** of the catawba rhododendron or mountain rosebay in the Great Smoky Mountains, USA, is now a tourist attraction.

- **The Dahurian** is a famous purply pink rhododendron from Siberia and Mongolia.

- **The Smirnow** was discovered in the 1880s high up in the Caucasus Mts on the Georgian–Turkish border.

- **Smirnows** have been bred with other rhododendrons to make them very hardy.

- **Azaleas** were once considered a separate group of plants, but they are now classified with rhododendrons.

◀ *Rhododendrons have thrived so well in places where they have been introduced that many people now consider them weeds.*

Carnivorous plants

- **Plants that trap** insects for food are called carnivorous plants. They live in places where they cannot get enough nitrogen from the soil and so the insects provide the nitrogen.

- **There are 550 species** living in places from the high peaks of New Zealand to the swamps of Carolina.

- **The butterwort** gets its name because its leaves ooze drops that make them glisten like butter. These drops contains the plant's digestive juices.

◄ *Pitchers hang on long tendrils that grow high into the branches of tropical rainforests. Some pitchers are tiny and can trap nothing bigger than an ant. The pitchers of the Nepenthes rajah are big enough for a rat.*

- **The sundew** can tell the difference between flesh and other substances and only reacts to flesh.

- **The sundew's** leaves are covered in tentacles that ooze a sticky substance called mucilage.

- **The sundew** wraps up its victims in its tentacles and suffocates them in slime in under ten seconds.

- **A Venus fly-trap's** trap will only shut if touched at least twice in 20 seconds.

- **Insects** are lured on to many carnivorous plants by sweet-tasting nectar – or the smell of rotting meat.

- **The juice** of a pitcher plant will dissolve a chunk of steak to nothing in a few days.

- **The bladders** of bladderworts were once thought to be air sacs to keep the plant afloat. In fact, they are tiny traps for water insects.

The fly touches hairs that send an electrical signal to cells on the side of the trap

◄ *Insects are lured into the jaw-like leaf trap of the Venus fly-trap with nectar. Once the insect lands, the jaws clamp shut on the victim in a fraction of a second. At once the plant secretes juices that drown, then dissolve, the insect.*

▼ *Like the Venus fly-trap, the Sarracenia is a native of North America. But instead of actively capturing its prey, it provides a deep tube for them to fall into. Insects drawn to the nectar round its rim slide in and are unable to climb out.*

When triggered, cells on the outside of the trap expand instantly and cells on the inside contract, pulling the trap shut

Tentacles covered in drops of sticky mucilage

◄ *When an insect lands on the sticky tentacles of a sundew, it struggles to free itself – but this struggling stimulates the tentacles to tighten their grip. Soon the tentacles exude a digestive juice that dissolves the victim.*

Temperate fruit

- **Fruits of temperate regions** must have a cool winter to grow properly.

- **The main temperate fruits** are apples, pears, plums, apricots, peaches, grapes and cherries.

- **Apples were eaten** by the earliest Europeans hundreds of thousands of years ago. They were spread through the USA by Indians, trappers and travellers like Johnny 'Appleseed' Chapman.

- **The world** picks 32 million tonnes of apples a year, half are eaten fresh and a quarter are made into the alcoholic drink cider. The USA is the world's leading producer of cider apples.

◀ *Plums are a kind of fruit called a drupe. This means the seed is contained inside a hard stone in the middle of the fruit.*

◀ *Pears are the second most important temperate fruit after apples. The leading producer is China.*

- **The world's most** popular pear is the Williams' Bon Chrétien or Bartlett. The best is said to to be the Doyenné du Comice, first grown in France in 1849.

- **New pear trees** are grown not from seeds but by grafting branches on to roots such as those of quinces.

- **Plums** came originally from the Caucasus Mountains in Turkey and Turkey is still the world's major plum grower. The damson plum came from Damascus.

- **Plums** are dried to make prunes.

- **The peach** is 87% water and has far fewer calories than fruit like apples and pears.

- **Grapes are grown** in vineyards to make wine. Grape-growing or viticulture is described in detail in Ancient Egyptian hieroglyphs of 2400BC.

Evergreen trees

> ★ STAR FACT ★
> The best-known evergreen is the Christmas tree – typically a Douglas fir or spruce.

- **An evergreen** is a plant that keeps its leaves in winter.

- **Many tropical broad-leaved trees** are evergreen.

- **In cool temperate regions** and the Arctic, most evergreen trees are conifers such as pines and firs. They have needle-like leaves.

- **Old needles** do turn yellow and drop, but they are replaced by new needles (unless the tree is unhealthy).

- **Evergreens** may suffer from sunscald – too much sun – in dry, sunny spots, especially in early spring.

- **Five coniferous groups,** including larches and cypresses, are not evergreen.

- **Many evergreens** were sacred to ancient cultures. The laurel or bay was sacred to the Greek god Apollo and used by the Romans as a symbol of high achievement.

- **Yews are grown** in many European churchyards – perhaps because the trees were planted on the sites by pagans in the days before Christianity. But the bark of the yew tree and its seeds are poisonous.

- **The sakaki** is sacred to the Japanese Shinto religion, and entire trees are uprooted to appear in processions.

▼ *In cool northern climates where the summers are brief, conifers stay evergreen to make the most of the available sunshine.*

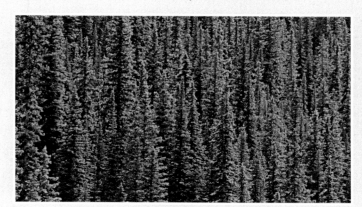

Fertilizers

▲ *Once the soil is broken up by ploughing, fertilizers are applied to prepare the soil for planting.*

- **Fertilizers** are natural or artificial substances added to soil to make crops and garden plants grow better.

- **Natural fertilizers** such as manure and compost have been used since the earliest days of farming.

- **Manure** comes mostly from farm animals, though in some countries, human waste is used.

- **Manure** has the chemicals nitrogen, phosphorus and potassium plants need for growth. It is also rich in humus, organic matter that helps keep water in the soil.

- **Artificial fertilizers** are usually liquid or powdered chemicals (or occasionally gas), containing a mix of nitrogen, phosphorus or potassium. They also have traces of sulphur, magnesium and calcium.

- **Nitrogen fertilizers**, also called nitrate fertilisers, are made from ammonia which is made from natural gas.

- **The first fertilizer** factory was set up by Sir John Lawes in Britain in 1843. He made superphosphate by dissolving bones in acid. Phosphates now come from bones or rocks.

- **Potassium fertilizers** come from potash dug up in mines.

- **The use of artificial** fertilizers has increased in the last 40 years, especially in the developed world.

- **Environmentalists** worry about the effects of nitrate fertilizers entering water supplies, and the huge amount of energy needed to make, transport and apply them.

Flower facts

- **The world's tallest** flower is the 2.5 m Titan arum which grows in the tropical jungles of Sumatra.

- **The Titan arum** is shaped so that flies are trapped in a chamber at the bottom.

- **The world's biggest flower** is Rafflesia, which grows in the jungles of Borneo and Sumatra. It is 1 m in diameter and weighs up to 11 kg.

- **Rafflesia** is a parasite and has no leaves, root or stems.

- **Rafflesia** and the Titan arum both smell like rotting meat to attract the insects that pollinate them.

▶ *Rafflesia was 'discovered' by British explorer John Arnold in 1818 and named by him after the famous British colonialist Stamford Raffles.*

- **The world's smallest flower** is the Wolffia duckweed of Australia. This is a floating water plant less than 0.6 mm across. It can only be seen clearly under a magnifying glass.

- **The biggest flowerhead** is the *Puya raimondii* bromeliad of Bolivia which can be up to 2.5 m across and 10 m tall and have 8000 individual blooms.

- **The Puya raimondii** takes 150 years to grow its first flower, then dies.

- **Two Australian orchids** bloom underground. No-one knows how they pollinate.

- **Stapelia flowers** not only smell like rotting meat to attract the flies that pollinate them – they look like it too (all pinky-brown and wrinkled).

Fruit

▶ There are three kinds of cherries – sweet, sour, and 'dukes', which are sweet-sour cross. We eat mainly sweet cherries like these.

- **Scientists** say a fruit is the ovary of a plant after the eggs are pollinated and grow into seeds. Corn grains, cucumbers, bean pods and acorns are fruit as well as apples and so on.

- **Some fruits** such as oranges are soft and juicy. The hard pips are the seeds.

- **With some fruits** such as hazelnuts and almonds, the flesh turns to a hard dry shell.

- **Fleshy fruits** are either berries like oranges which are all flesh, aggregate fruits like blackberries which are made from lots of berries from a single flower or multiple fruits like pineapples which are single fruits made from an entire multiple flowerhead.

- **Legumes** such as peas and beans are soft, dry fruits held in a case called a pod.

- **Berries** and other juicy fruits are called 'true fruits' because they are made from the ovary of the flower alone.

- **Apples and pears** are called 'false fruits' because they include parts other than the flower's ovary.

- **In an apple** only the core is the ovary.

- **Drupes** are fruit like plums, mangoes and cherries with no pips but just a hard stone in the centre containing the seeds. Aggregate fruits like raspberries are clusters of drupes.

- **Walnuts and dogwood** are actually drupes like cherries.

Roots

- **Roots are** the parts of a plant that grow down into soil or water, anchoring it and soaking up all the water and minerals the plant needs to grow.

- **In some plants** such as beetroots, the roots are also a food store.

- **When a seed** begins to grow, its first root is called a primary root. This branches into secondary roots.

> ★ STAR FACT ★
> The roots of the South African wild fig tree can grow 120 m down into the ground.

- **Roots** are protected at the end by a thimble-shaped root cap as they probe through the soil.

- **On every root** there are tiny root hairs that help it take up water and minerals.

- **Some plants,** such as carrots, have a single large root, called a taproot, with just a few fine roots branching off.

- **Some plants** such as grass have lots of small roots, called fibrous roots, branching off in all directions.

- **Some kinds of orchid** that live on trees have 'aerial' roots that cling to the branches.

- **Mistletoe** has roots that penetrate its host tree.

◀ A tree blown over in a gale reveals some of the dense mat of roots it needs to get enough water and nutrients.

Harvesting grain

- **When grain** is ripe it is cut from its stalks. This is called reaping.
- **After reaping** the grain must be separated from the stalks and chaff (waste). This is called threshing.
- **After threshing** the grain must be cleaned and separated from the husks. This is called winnowing.
- **In some places** grain is still reaped in the ancient way with a long curved blade called a sickle.
- **In most developed countries** wheat and other cereals are usually harvested with a combine harvester.
- **A combine harvester** is a machine that reaps the grain, threshes it, cleans it and pours it into bags or reservoirs.
- **The first horse-drawn** combine was used in Michigan in 1836, but modern self-propelled harvesters only came into use in the 1940s.
- **If the grain is damp** it must be dried immediately after harvesting so it does not rot. This is always true of rice.

▲ *Combine harvesters driven by a single man have replaced the huge teams of people with sickles of ancient times.*

- **If the grain is too damp** to harvest, a machine called a windrower may cut the stalks and lay them in rows to dry in the wind for later threshing and cleaning.
- **A successful harvest** is traditionally celebrated with a harvest festival. The cailleac or last sheaf of corn is said to be the spirit of the field. It is made into a harvest doll, drenched with water and saved for the spring planting.

Desert plants

▲ *Surprisingly many plants can survive the dryness of deserts, including cactuses and sagebushes.*

- **Some plants** find water in the dry desert with very long roots. The Mesquite has roots that can go down as much as 50 m deep.
- **Most desert plants** have tough waxy leaves to cut down on water loss. They also have very few leaves; cactuses have none at all.

- **Pebble plants** avoid the desert heat by growing partly underground.
- **Window plants** grow almost entirely underground. A long cigar shape pokes into the ground, with just a small green window on the surface to catch sunlight.
- **Some mosses and lichens** get water by soaking up dew.
- **Resurrection trees** get their name because their leaves look shrivelled brown and dead most of the time – then suddenly turn green when it rains.
- **The rose of Jericho** is a resurrection plant that forms a dry ball that lasts for years and opens only when damp.
- **Daisies** are found in most deserts.
- **Cactuses and ice plants** can store water for many months in special water storage organs.

★ **STAR FACT** ★
The quiver tree drops its branches to save water in times of drought.

Fungi

★ STAR FACT ★
Many mould fungi are the source of life-saving antibiotic drugs such as penicillin .

- **Fungi** are a huge group of 50,000 species. They include mushrooms, toadstools, mould, mildew and yeast.

- **Fungi** are not plants, because they have no chlorophyll to make their food. So scientists put them in a group or kingdom of their own.

- **Because fungi** cannot make their own food, they must live off other plants and animals – sometimes as partners, sometimes as parasites.

- **Parasitic fungi** feed off living organisms; fungi that live off dead plants and animals are called saprophytic.

- **Fungi** feed by releasing chemicals called enzymes to break down chemicals in their host. The fungi then use the chemicals as food.

▶ These are some of the tens of thousands of different fungi, which are found growing everywhere from rotting tree stumps to inside your body.

- **Cheeses** like Camembert, Rocquefort, Stilton and Danish Blue get their distinctive flavours from chemicals made by moulds added to them to help them ripen. The blue streaks in some cheeses are actually moulds.

- **Fungi are made** of countless cotton-like threads called hyphae which absorb the chemicals they feed on. Hyphae are usually spread out in a tangled mass. But they can bundle together to form fruiting bodies like mushrooms.

- **Some fungi** grow by spreading their hyphae in a mat or mycelium; others scatter their spores. Those that grow from spores go through the same two stages as mosses.

- **Truffles** are fungi that grow near oak and hazel roots. They are prized for their flavour and sniffed out by dogs or pigs. The best come from Perigord in France.

The field mushroom, grown wild or cultivated, is the mushroom most widely eaten

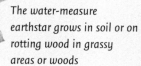

Honey mushrooms belong to the Armillaria genus of fungi, which includes the world's largest and oldest living organisms

Fungi can grow in all kinds of shapes, earning them names like this orange peel fungi

The destroying angel is the most poisonous of all fungi, and usually kills anyone who eats one

The water-measure earthstar grows in soil or on rotting wood in grassy areas or woods

Fly agaric is a toadstool – that is, a poisonous mushroom. It is easy to recognize from its spotted red cap

The chanterelle is a sweet-smelling, edible amber-coloured mushroom. But it looks very like the poisonous jack o'lantern

Puffballs have big round fruiting bodies that dry out and puff out their spores in all directions when burst

Heathlands

▲ *Heather and other heathland plants are usually pollinated by bees and birds like sunbirds.*

- **Heathland** goes under many different names, including scrubland, shrubland and chaparral.

- **Heathlands** occur where the soil is too dry or too poor for trees to grow – typically in Mediterranean regions or areas of sandy soil.

- **Many heathlands** are not natural, but places where human activity has so changed the environment that trees can no longer grow.

- **The most common** heathland shrub is heather. Underneath grasses, sedges and flowers like daisies and orchids grow.

- **Many heathland shrubs** like gorse are thorny to stop animals eating them.

- **The maquis** are the heathlands of the Mediterranean, dominated by tough evergreen shrubs and small trees.

- **Many maquis** plants are aromatic (have a strong scent) – such as mints, laurels and myrtles.

- **Spring blossoms** in the mallee heaths of Australia are so spectacular that they are a tourist attraction.

- **Mallee** is a kind of eucalyptus tree typical of the area.

- **Chaparral** is heathland in California. The climate is Mediterranean, with mild winters and warm summers. The main plants are sages and small evergreen oaks.

Timber

- **Timber** is useful wood. Lumber is a North American term for timber once it is sawn or split.

- **Lumberjacks** are people who cut down trees using power saws or chainsaws.

- **Round timbers** are basically tree trunks that have been stripped of their bark and branches and cut into logs.

- **Round timbers** are used for fencing and telegraph poles or driven into the ground as 'piles' to support buildings and quays.

- **Lumber** is boards and planks sawn from logs at sawmills. At least half of lumber is used for building.

▲ *Tree surgeons stripping branches from a felled tree with chainsaws.*

- **Before lumber** can be used, it must usually be seasoned (dried) or it will shrink or twist. Sometimes it is dried in the open air, but more often it is warmed in a kiln or treated with chemicals.

- **Sometimes** planks are cut into thin slices called veneers.

- **Plywood** is three or more veneers glued together to make cheap, strong wood. Chipboard is wood chippings and sawdust mixed with glue and pressed into sheets.

- **Softwod lumbers** come from trees such as pines, larches, firs, hemlocks, redwoods and cedars.

- **MDF** or medium density fibreboard is made from glued wood fibres.

Symbiosis

- **Living things** that feed off other living things are called parasites.

- **Living things** that depend on each other to live are called symbiotic.

- **Many tropical rainforest trees** have a symbiotic relationship with fungi on their roots. The fungi get

▼ *Leaf-cutter ants cut up leaves and line their nests – not for themselves, but for the fungi which grow on the leaves. The ants eat the fungi.*

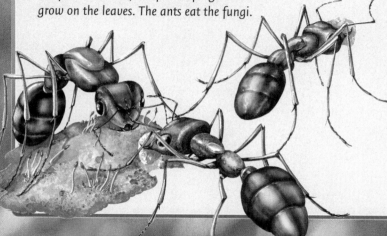

energy from the trees and in return give the trees phosphorus and other nutrients.

- **A phyte is a plant** that grows on another plant.

- **Epiphytes** are plants that grow high up on other plants, especially in tropical rainforests (see epiphytes).

- **Saprophytes** are plants and fungi that depend on decomposing material, not sunlight, for sustenance.

- **Most orchids** are saprophytic as seedlings.

- **Corsiaceae orchids** of New Guinea, Australia and Chile are saprophytic all their lives.

- **Various ants,** such as leaf-cutter and harvester ants in tropical forests, line their nests with leaves which they cut up. The leaves provide food for fungi which, in turn, provide food for the ants.

Tree flowers

- **All trees have flowers,** but the flowers of conifers are usually tiny compared with those of broad-leaved trees.

- **Flowers** are a tree's reproductive organs.

- **Some flowers are male.** Some are female.

- **Sometimes the male** and female flowers are on separate trees. Sometimes, as in willows and some conifers, they are on the same tree.

- **'Perfect' flowers** like those of cherry and maple trees have both male and female parts.

- **Pollen** is carried from male flowers by insects or the wind to fertilize female flowers.

- **A blossom** can be any flower, but often refers especially to the beautiful flowers of fruit trees such as cherries and apples in spring.

- **Many blossoms** are pink and get their colour from what are called anthocyanin pigments – the same chemical colours that turn leaves red in autumn.

- **Washington DC** is famous for its Cherry Blossom Festival each spring.

- **Omiya** in Japan is famous for its park full of cherry trees which blossom in spring.

▼ *Apple blossoms are usually pink. They bloom quite late in spring, after both peach and cherry blossoms.*

Parts of a plant

All plants that grow from seeds have flowers, although not all are as bright and colourful as these

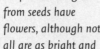

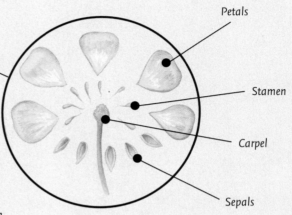

Petals

Stamen

Carpel

Sepals

Flowers usually open only for a short time. Before they open, they are hidden in tight green buds

The leaves are the plant's powerhouses, using sunlight to make sugar, the plant's fuel

◄ Plants come in many shapes and sizes from tiny wildflowers to giant trees 100 m tall. But they all tend to have the same basic features – roots, stem, leaves and flowers.

The stem supports the leaves and flowers and channels water and minerals up from the roots

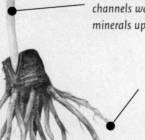

The roots grow down into soil or water. They hold the plant in place, and allow it to draw up water and minerals

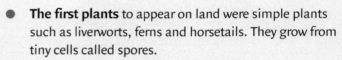

- **The first plants** to appear on land were simple plants such as liverworts, ferns and horsetails. They grow from tiny cells called spores.

- **Today, most plants** grow not from spores but from seeds. Unlike primitive plants, seed-making plants have stems, leaves and often roots and flowers.

- **The stem of a plant** supports the leaves and flowers. It also carries water, minerals and food up and down between the plant's leaves and roots.

- **A terminal bud** forms the tip of each stem. The plant grows taller here.

- **Lateral buds** grow further back down the stem at places called nodes.

- **Some lateral buds** develop into new branches. Others develop into leaves or flowers.

- **The leaves** are the plant's green surfaces for catching sunlight. They use the sun's energy for joining water with carbon dioxide from the air to make the sugar the plant needs to grow (see photosynthesis).

- **The roots** are the parts of the plant that grow down into soil or water. They anchor the plant in the ground and soak up all the water and minerals it needs to grow.

- **The flowers** are the plant's reproductive organs. In gymnosperms – conifers, cycads and gingkos – the flowers are often small and hidden. In angiosperms (flowering plants) they are usually much more obvious.

Tropical trees

◀ Mangrove trees are famous for their dangling pods which can drop like a sword on passersby.

- **Nearly all tropical trees** are broad-leaved trees.

- **Most tropical trees** are evergreen. Only where there is a marked dry season in 'monsoon' regions do some trees loose their leaves to save water.

- **Most tropical trees** are slow-growing hardwoods such as teak and mahogany. Once cut down, they take many years to replace.

- **Mahogany** is a tall evergreen tree with beautiful hard wood that turns red when it matures after a century or so.

- **Most mahogany** wood comes from trees such as the African *Khaya* or the *Shorea* from the Philippines.

- **The best mahogany** is from the tropical American *Swietenia macrophylla*.

- **Balsa** is so light and such a good insulator that it is used to make passenger compartments in aircraft.

- **Teak** is a deciduous tree from India. It is one of the toughest of all woods and has been used to construct ships and buildings for more than 2000 years.

- **Chicle** is a gum drained from the Central American sapota tree in the rainy season. It is the main ingredient in chewing-gum. The best comes from Guatemala.

Dandelions and daisies

- **Dandelions and daisies** are both members of a vast family called *Asteraceae*.

- **All *Asteraceae*** have flower heads with many small flowers called florets surrounded by leaf-like structures called bracts.

- **There are over 20,000** different *Asteraceae*.

- **Garden *Asteraceae*** include asters, dahlias and chrysanthemums.

- **Wild *Asteraceae*** include burdock, butterbur and ragweed, thistles and sagebrush.

▶ Daisies look like a single bloom, but they actually consist of many small flowers. Those around the edge each have a single petal.

- **Lettuces, artichokes** and sunflowers are all *Asteraceae*.

- **The thistle** is the national emblem of Scotland.

- **Dandelions** are bright yellow flowers that came originally from Europe, but were taken to America by colonists. Unusually, their ovaries form fertile seeds without having to be pollinated, so they spread rapidly.

- **The name dandelion** comes from the French *dent de lion*, meaning lion's tooth – because its leaves have edges that look like sharp teeth.

- **The daisy** gets its name from the Old English words 'day's eye' – because like an eye its blooms open in the day and close at night.

Berries

- **Berries** are fleshy fruit which contain lots of seeds. The bright colours attract birds which eat the flesh. The seeds pass out in the birds' droppings and so spread.

- **Bananas**, tomatoes and cranberries are all berries.

- **Strawberries,** raspberries and blackberries are not true berries. They are called 'aggregate' fruits because each is made from groups of tiny fruit with one seed.

- **Gean, damson and blackthorn berries** contain a single seed. Holly berries and elderberries contain many.

- **Cloudberries** are aggregate fruits like raspberries. The tiny amber berries grow close to the ground in the far north, and are collected by Inuits and Sami people in autumn to freeze for winter food.

◄ *Most berries are shiny bright red to attract birds.*

- **Cloudberries** are also known as salmonberries, bakeberries, malka and baked appleberries.

- **Cranberries** grow wild on small trailing plants in marshes, but are now cultivated extensively in the USA in places such as Massachusetts.

- **Wild huckleberries** are the American version of the European bilberry. But the evergreen huckleberry sold in florists is actually a blueberry.

- **The strawberry tree's** Latin name is *unedo*, which means 'I eat one'. The red berries are not as tasty as they look.

- **A Greek myth** tells how the wine-red mulberry was once white but was stained red by the blood of the tragic lovers Pyramus and Thisbe, whose story is retold in Shakespeare's *Midsummer Night's Dream*.

Ash trees

- **Ash trees** are 70 species of deciduous trees that grow through much of northern Eurasia and North America.

- **Ash trees** are among the most beautiful of all trees and are prized for their wood. It was once used to make oars and handles for axes and tennis rackets and for skis.

- **The tallest of all flowering plants** is the Australian mountain ash which grows over 100 m tall.

◄ *The leaves of the ash grow opposite each other in groups of five to nine and have tooth edges. The clusters of flowers are small and often showy.*

- **Ash trees** are part of the olive family.

- **The Vikings** worshipped the ash as a sacred tree. Yggdrasil, the Tree of the World, was a giant ash whose roots reached into hell but whose crown reached heaven.

- **In Viking myth** Odin, the greatest of the gods, created the first man out of a piece of ash wood.

- **The manna ash** got its name because it was once thought that its sugary gum was manna. Manna was the miraculous food that fell from heaven to feed the Biblical Children of Israel in the desert as they fled from Egypt.

- **The mountain ash** is also known as the rowan or quickbeam. In America it is known as dogberry. It is not related to other ash trees.

- **Rowan trees** were once linked to witchcraft. The name may come from the Viking word *runa*, meaning charm. Rowan trees were planted in churchyards, and the berries were hung over doors on May Day, to ward off evil.

- **Rowans** grow higher up mountains than any other tree.

Dicotyledons

▲▶ *Dicots, like the Japanese maple above, all begin life as a pair of leaves growing from a seed, like those on the right.*

- **Dicotyledons** are one of two basic classes of flowering plant. The other is monocotyledons.

- **Dicotyledons** are also known as dicots or Magnoliopsida.

- **Dicots** are plants that sprout two leaves from their seeds.

  - **There are about 175,000** dicots – over three-quarters of all flowering plants.

  - **Dicots** include most garden plants, shrubs and trees as well as flowers such as magnolias, roses, geraniums and hollyhocks.

  - **Dicots** grow slowly and at least 50% have woody stems.

- **The flowers** of dicots have sets of four or five petals.

- **Most dicots** have branching stems and a single main root called a taproot.

- **The leaves of dicots** usually have a network of veins rather than parallel veins.

- **Dicots** usually have a layer of ever-growing cells near the outside of the stem called the cambium.

Orchids

★ **STAR FACT** ★
To attract male bees, the bee orchid has a lip that looks just like a female bee.

- **Orchids** are a group of over 20,000 species of flower, growing on every continent but Antarctica.

- **In the moist tropics** many grow on the trunks and branches of trees and so are called epiphytes.

- **A few,** such as the Bird's nest orchid, are saprophytes, living off rotting plants in places where there is no light.

- **Some species** are found throughout the tropics, such as *Ionopsis utricularioides*. Others grow on just a single mountain in the world.

- **Orchids** have a big central petal called the lip or labellum. It is often shaped like a cup, trumpet or bag.

- **The fly orchid** of Ecuador has a lip shaped like a female tachinid fly to attract male flies.

- **The flavour vanilla** comes from the vanilla orchid.

- **Ancient Greek** couples expecting a baby often ate the roots of the early purple orchid. They believed that if the man ate the flower's large root the baby would be a boy. If the woman ate the small root, the baby would be a girl.

- **In Shakespeare's** *Hamlet*, the drowned Ophelia is covered in flowers, including the early purple orchid, famous as a love potion. Hamlet's mother says that 'cold maids' call the flowers 'dead men's fingers'.

▶*The early purple orchid was said to have grown beneath Christ's cross and the red spots on its leaves were said to be left by falling drops of Christ's blood.*

Cycads and gingkos

- **Cycadophytes and gingkophytes** were the first seed plants to appear on land. The cycads and gingkos of today are their direct descendants.

- **Like conifers,** cycads and gingkos are gymnosperms. This means their seeds do not develop inside a fruit like those of flowering plants or angiosperms.

- **Cycads** are mostly short, stubby, palm-like trees. Some are many thousands of years old.

- **Cycads have** fern-like leaves growing in a circle round the end of the stem. New leaves sprout each year and last for several years.

- **The gingko** is a tall tree that comes from China.

- **The gingko** is the only living gingkophyte.

- **The gingko** is the world's oldest living seed-plant.

- **Fossil leaves** identical to today's gingko have been found all over the world in rocks formed in the Jurassic period, 208–144 million years ago.

▲ *The gingko is a remarkable living fossil – the only living representative of the world's most ancient seed plants.*

- **Scientifically,** the gingko is called *Gingko Biloba*. It is also called the maidenhair tree.

- **All today's** gingkos may be descended from trees first cultivated in Chinese temple gardens 3000 years ago.

River plants

▲ *Water crowfoots are buttercups that grow in water. They may have both round floating leaves like these and feathery submerged leaves.*

- **Some aquatic (water) plants** are rooted in the mud and have their leaves above the surface like water lilies.

- **Some water plants** grow underwater but for their flowers, like water milfoils and some plantains. They may have bladders or air pockets to help keep the stem upright.

- **Tiny plants** called algae grow in red, green or brown films on rivers, lakes and swamps.

- **Water hyacinths** are purple American water flowers. They grow quickly and can clog up slow streams.

- **Giant water lilies** have huge leaves with the edges upturned like a shallow pan to keep them afloat.

- **The leaves** of the royal or Amazon lily can be 2 m across.

- **Papyrus** is a tall, grass-like water plant that grows in the Nile river. Stems were rolled flat by the Ancient Egyptians to write on. The word 'paper' comes from papyrus.

- **Many grass-like** plants grow in water, including reeds, mace, flag and rushes such as bulrushes and cattails.

- **Mangroves, bald cypresses,** cotton gum and other 'hydrophytic' trees are adapted to living in water.

> ★ STAR FACT ★
> Mexico's largest lake, Lake Chapala, is sometimes choked with water hyacinths.

Development of a flower

- **Flowers have both** male parts, called stamens, and female parts, called carpels. Seeds for new plants are made when pollen from the stamens meets the flower's eggs inside the carpels.

- **The carpel** contains the ovaries, where the flower's eggs are made. It is typically the short thick stalk in the middle of the flower.

- **A flower** may have just one carpel or several joined together. Together, they are called the pistil.

- **The stamens** make pollen. They are typically spindly stalks surrounding the carpels.

- **Pollen is made** in the anthers on top of the stamens.

- **Pollen** is trapped on the top of the ovary by sticky stigma.

▶ Most flowers rely on bees and butterflies to fertilize them by transferring pollen from the stamens to the carpels. So, like this orchid, flowers have developed wonderful colours and scents to attract the insects to them.

- **Pollen** is carried down to the ovary from the stigma via a structure called the style. In the ovary it meets the eggs and fertilizes them to create seeds.

- **Before the flower opens,** the bud is enclosed in a tight green ball called the calyx. This is made up from tiny green flaps called sepals.

- **The colourful part of the flower** is made from groups of petals. The petals make up what is called the corolla. Together the calyx and the corolla make up the whole flower head, which is called the perianth. If petals and sepals are the same colour, they are said to be tepals.

> ★ STAR FACT ★
> A 'perfect' flower is one which has both
> stamens and carpels; many have one missing.

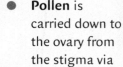

1. The fully formed flower is packed away inside a bud. Green flaps or sepals wrap tightly round it

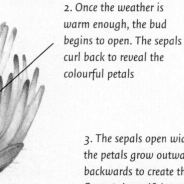

2. Once the weather is warm enough, the bud begins to open. The sepals curl back to reveal the colourful petals

3. The sepals open wider and the petals grow outwards and backwards to create the flower's beautiful corolla

▶ At the right time of year, buds begin to open to reveal flowers' blooms so that the reproductive process can begin. Some flowers last just a day or so. Others stay blooming for months on end before the eggs are fertilized, and grow into seeds.

4. The flower opens fully to reveal its bright array of pollen sacs

Deciduous trees

◀ *In autumn, the leaves of deciduous trees turn glorious browns, reds and golds and then drop off. New leaves grow in the spring.*

- **Deciduous trees** are trees that lose their leaves once a year.

- **In cool places,** deciduous trees lose their leaves in autumn to cut their need for water in winter when water may be frozen.

- **In the tropics** deciduous trees lose their leaves at the start of the dry season.

- **Leaves fall** because a layer of cork grows across the leaf stalk, gradually cutting off its water supply.

- **Eventually the leaf** is only hanging on by its veins, and is easily blown off by the wind.

- **Leaves go brown** and other colours in autumn because their green chlorophyll breaks down, letting other pigments shine through instead.

- **Among the most spectacular** autumn colours are those of the sweet gum, brought to Europe from Mexico c.1570.

- **The main deciduous trees** in cool climates are oaks, beeches, birches, chestnuts, aspens, elms, maples and lindens.

- **Most deciduous trees** are broad-leaved, but five conifer groups including larches are deciduous.

- **Some tropical evergreen trees** are deciduous in regions where there is a marked dry season.

Tree leaves

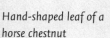

- **Trees** can be divided into two groups according to their leaves: broad-leaved trees and conifers with needle-like leaves.

- **The leaves** of broad-leaved trees are all wide and flat to catch the sun, but they vary widely in shape.

- **You can identify** trees by their leaves. Features to look for are not only the overall shape, but also: the number of leaflets on the same stalk, whether leaflets are paired or offset and if there are teeth round the edges of the leaves.

- **Trees such as birches** and poplars have small triangular or 'deltoid' leaves; aspens and alders have round leaves.

- **Limes** and Indian bean trees have heart-shaped or 'cordate' leaves.

Hand-shaped leaf of a horse chestnut

Long, narrow willow leaves

- **Maples** and sycamores have leaves shaped a bit like hands, which is why they are called 'palmate'.

- **Ash and walnut trees** have lots of leaflets on the same stalk, giving them a feathery or 'pinnate' look.

- **Oaks and whitebeams** have leaves indented with lobes round the edge.

- **Many shrubs,** like magnolias and buddleias, and trees like willows, cherries, sweet chestnuts and cork oaks, have long narrow leaves.

- **Elms, beeches,** pears, alders and many others have oval leaves.

Lobed leaves of the English oak

Pinnate or feather-shaped walnut leaves

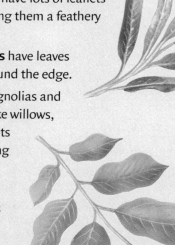

Root vegetables

▲▶ *Potatoes and carrots are important root vegetables. Carrot is a source of vitamin A, potatoes are sources of many vitamins, such as C.*

- **Vegetables** are basically any part of a plant eaten cooked or raw, except for the fruit.

- **Root vegetables** are parts of a plant that grow underground in the soil.

- **Turnips, rutabaga,** beets, carrots, parsnips and sweet potatoes are the actual roots of the plant.

- **Potatoes and cassava** are tubers or storage stems.

- **Potatoes** were grown in South America at least 1800 years ago. They were brought to Europe by the Spanish in the 16th century.

- **Poor Irish** farmers came to depend on the potato, and when blight ruined the crop in the 1840s, many starved.

- **Yams are tropical roots** similar to sweet potatoes. They are an important food in West Africa. A single yam can weigh 45 kg or more.

- **Mangel-wurzels** are beet plants grown mainly to feed to farm animals.

- **Tapioca** is a starchy food made from cassava that once made popular puddings.

- **Carrots came** originally from Afghanistan, but were spread around the Mediterranean 4000 years ago. They reached China by the 13th century AD.

Poisonous fungi

- **Many fungi** produce poisons. Scientists call poisons 'toxins' and poisons made by fungi 'mycotoxins'.

- **Some poisonous fungi** are very small microfungi which often form moulds or mildew. Many are either 'sac' fungi (*Ascomycetes*) or 'imperfect' fungi (*Deuteromycetes*).

- **Ergot** is a disease of cereals, especially rye, caused by the sac fungus *Claviceps purpurea*. If humans eat ergot-infected rye, they may suffer an illness called St Anthony's Fire. Ergot is also the source of the drug LSD.

- *Aspergillus* is an imperfect fungus that may cause liver damage or even cancer in humans.

- **False morel** is a poisonous sac fungus as big as a mushroom. True morels are harmless.

- **About 75 kinds** of mushroom are toxic to humans and so called toadstools. Most belong to the Amanita family, including destroying angels, death caps and fly agarics.

- **Death caps** contain deadly phalline toxins that kill most people who eat the fungus.

- **Fly agaric** was once used as fly poison.

- **Fly agaric** and the *Psilocybe mexicana* mushroom were eaten by Latin American Indians because they gave hallucinations.

> ★ STAR FACT ★
> Athlete's foot is a nasty foot condition caused by a fungus.

▶ *Fly agaric contains the poison muscarine. It rarely kills but makes you sick and agitated.*

Broad-leaved woodlands

▲ *Avenues of broad-leaved trees form shady paths in summer but are light in winter when the trees are bare.*

- **Forests** of broad-leaved, deciduous trees grow in temperate regions where there are warm, wet summers and cold winters – in places like North America, western Europe and eastern Asia.

- **Broad-leaved deciduous** woods grow where temperatures average above 10°C for over six months a year, and the average annual rainfall is over 400 mm.

▼ *Plenty of light can filter down through deciduous trees – especially in winter when the leaves are gone – so all kinds of bushes and flowers grow in the woods, often blooming in spring while the leaves are still thin.*

- **If there are** 100 to 200 days a year warm enough for growth, the main trees in broad-leaved deciduous forests are oaks, elms, birches, maples, beeches, aspens, chestnuts and lindens (basswood).

- **In the tropics** where there is plenty of rainfall, broad-leaved evergreens form tropical rainforests.

- **In moist western Europe,** beech trees dominate woods on well-drained, shallow soils, especially chalkland; oak trees prefer deep clay soils. Alders grow in waterlogged places.

- **In drier eastern Europe,** beeches are replaced by durmast oak and hornbeam and in Russia by lindens.

- **In American woods**, beech and linden are rarer than in Europe, but oaks, hickories and maples are more common.

- **In the Appalachians** buckeye and tulip trees dominate.

- **There is a wide range** of shrubs under the trees including dogwood, holly, magnolia, as well as woodland flowers.

> ★ **STAR FACT** ★
> Very few woods in Europe are entirely natural; most are 'secondary' woods, growing on land once cleared for farms.

Cut flowers

◄ Holland is still famous for its flower markets.

- **Cut flowers** are flowers sold by the bunch in florists.

- **The cut flower** trade began in the Netherlands with tulips in the 1600s.

- **In 1995** 60% of the world's cut flowers were grown in Holland.

- **Latin American countries** like Columbia, Ecuador, Guatemala and Costa Rica are now major flower-growers. So too are African countries like Kenya, Zimbabwe, South Africa, Zambia and Tanzania.

- **In China** the growing popularity of St Valentine's day has meant huge areas of China are now planted with flowers.

- **After cutting,** flowers are chilled and sent by air to arrive in places like Europe and North America fresh.

- **Most of the world's** cut flowers are sold through the huge flower market in Rotterdam in Holland.

- **By encouraging** certain flowers, flower-growers have made cut flowers last longer in the vase – but they have lost the rich scents they once had. Scientists are now trying to reintroduce scent genes to flowers.

- **A corsage** is a small bouquet women began to wear on their waists in the 18th century.

- **A nosegay** was a small bouquet Victorian ladies carried in their hands. If a man gave a lady a red tulip it meant he loved her. If she gave him back a sprig of dogwood it mean she didn't care. Various pink flowers meant 'no'.

Wheat

> ★ STAR FACT ★
> The world grows enough wheat a year to fill a line of trucks stretching a quarter of the way to the Moon.

- **Wheat grows** over more farmland than any other crop and is the basic food for 35% of the world's population.

- **Wheat was** one of the first crops ever grown, planted by the first farmers some 11,000 years ago.

- **Today** there are over 30 varieties. Among the oldest are emmer and einkorn.

- **Spring wheat** is planted in spring and then harvested in early autumn.

- **Winter wheat** is planted in autumn and harvested the following summer.

- **Wheat** is a kind of grass, along with other cereals.

- **Young wheat plants** are short and green and look like ordinary grass, but as they ripen they turn golden and grow between 0.6 and 1.5 m tall.

- **Branching from the main stem** are stalks called tillers. Wrapped round them is the base or sheath of the leaves. The flat top of the leaf is called the blade.

- **The head of the corn** where the seeds or grain grow is called the ear or spike. We eat the seed's kernels (core), ground into flour to make bread, pasta and other things.

▼ An ear of wheat with the seeds which are stripped of their shells or husks before being ground to make flour.

Annuals and biennials

▲ *Foxgloves are typical biennials, flowering in their second summer, then dying back.*

- **Annuals** are plants that grow from seed, flower, disperse their seeds and die in a single season.

- **Some annuals' seeds** lie dormant in the ground before conditions are right for germination.

- **With an annual,** forming flowers, fruits and seeds exhausts the plant's food reserves so the green parts die.

- **Many crops** are annuals, including peas and beans, squashes, and cereals such as maize and wheat.

- **Annual flowers** include petunias, lobelias, buttercups and delphiniums.

- **Biennials** live for two years.

- **In the first year** the young plant grows a ring of leaves and builds up an underground food store such as a bulb or taproot like beetroots and carrots. The food store sustains the plant through the winter.

- **In the second year** the plant sends up a stem in spring. It flowers in the summer.

- **Many vegetables** are biennials, including beetroot, carrots and turnips.

- **Biennial flowers** include wallflowers, carnations, sweet williams and evening primroses.

Bamboo

- **Bamboos** are giant, fast-growing grasses with woody stems.

- **Most bamboos** grow in east and southeast Asia and on islands in the Indian and Pacific oceans.

- **Bamboo stems** are called culms. They often form dense thickets that exclude every other plant.

- **Bamboo culms** can reach up to 40 m and grow very fast. Some bamboos grow 1 m every three days.

- **Most bamboos** only flower every 12 years or so. Some flower only 30–60 years. *Phyllostachys bambusoides* flowers only after 120 years.

- **Pandas** depend on the *Phyllostachys* bamboo, and after it flowers they lose their source of food.

▲ *Bamboo looks like trees with its tall woody stems and big leaves, but it is actually grass.*

- **The flowering** of the muli bamboo around the Bay of Bengal every 30–35 years brings disaster as rats multiply to take advantage of the fruit.

- **The Chinese** have used the hollow stems of bamboo to make flutes since before the Stone Age. The Australian aboriginals use them to make droning pipes called didgeridoos.

- **Bamboo** is an incredibly light, strong material, and between 1904 and 1957 athletes used it for pole-vaulting. American Cornelius Warmerdam vaulted 4.77 m with a bamboo pole.

- **Bamboo** has long been used to make paper. The Bamboo Annals, written on bamboo, are the oldest written Chinese records, dating from the 8th century BC.

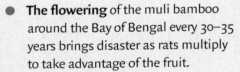

Citrus fruit

- **Citrus fruits** are a group of juicy soft fruits covered with a very thick, waxy, evenly coloured skin in yellow, orange or green.

- **Citrus** fruits include lemons, limes, oranges, tangerines, grapefruits and shaddocks.

▼ *Orange trees are planted in groves. The fruit are green when they first appear, but turn orange as they ripen.*

> ★ STAR FACT ★
> Citrus fruits are richer in Vitamin C than any other fruit or vegetable.

- **Inside the skin,** the flesh of a citrus fruit is divided into clear segments, each usually containing one or several seeds or pips.

- **Citrus fruits** grow in warm Mediterranean climates, and they are very vulnerable to frost.

- **Some citrus** fruit-growers warm the trees with special burners in winter to avoid frost-damage.

- **The sharp tang** of citrus fruits comes from citric acid.

- **Lemons** were spread through Europe by the crusaders who found them growing in Palestine.

- **Columbus** took limes to the Americas in 1493.

- **Scottish physician** James Lind (1716-1794) helped eradicate the disease scurvy from the British navy by recommending that sailors eat oranges and lemons.

Parasites

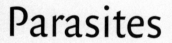

- **Parasitic plants** are plants that get their food not by using sunlight but from other plants, at the others' expense.

- **In the gloom of** tropical rainforests, where sunlight cannot penetrate, there are many parasitic plants growing on the trees.

- **Lianas** save themselves energy growing a trunk by climbing up other trees, clinging on with little hooks.

- **Rafflesia,** the world's biggest flower, is a parasite that feeds on the roots of lianas.

- **Figs** begin growing from seeds left high on branches by birds or fruit bats.

- **Fig roots** grow down to the ground around the tree, strangling it by taking its water supply. The tree then dies away, leaving the fig roots as a hollow 'trunk'.

- **Mistletoes** are semi-parasitic plants that wind round trees. They draw some of their food from the tree and some from sunlight with their own leaves.

- **Viscum album** mistletoe was held sacred by Druids 2000 years ago.

- **The druid** belief in the magic power of mistletoe survives in the tradition of kissing under the mistletoe at Christmas.

- **Broomrapes** grow on sugarcane roots; witchweeds grow on maize and rice roots.

◄ *Mistletoe, with its distinctive white berries, grows on apple and poplar trees in Eurasia and oaks in America.*

Epiphytes

- **Epiphytes** are plants that grow high above the ground in tropical rainforests, on tree branches.

- **Epiphytes** are often known as air plants because they seem to live on air – attached neither to the ground nor to any obvious source of nutrients.

- **Epiphytes** get their water and minerals from rain water, and from debris on the branch.

- **Various** orchids, ferns and bromeliads are epiphytes in tropical forests.

- **There are also epiphytes** in cooler places, including lichens, mosses, liverworts and algae.

- **Bromeliads** belong to a big family of plants called the pineapple family. At least half of them are epiphytes.

- **The pineapple fruit** is the best-known bromeliad.

- **All but one bromeliad** come from America, but they live in a huge range of habitats, living anywhere from on cacti in deserts to moist forests high up mountains.

▲ Trees in tropical rainforests are often covered in epiphytes festooned on every bough and branch.

- **The smallest bromeliads** are moss-like *Tillandsia bryoides*, just a few centimetres long.

- **The biggest bromeliad** is *Puya raimondii*, with a stem up to 4 m long and a flower over 4 m tall.

Seeds and nuts

- **Seeds are the tiny** hard capsules from which most new plants grow.

- **Seeds** develop from the plant's egg once it is fertilized by pollen.

- **Each seed** contains the new plant in embryo form plus a store of food to feed it until it grows leaves.

◀ *Neither Brazil nuts nor coconuts are true nuts. Coconuts (right) are not true nuts, but the stones of drupes. Brazil nuts (left) are just large seeds.*

- **The seed** is wrapped in a hard shell or testa.

- **Some fruit** contain many seeds; nuts are fruit with a single seed in which the outside has gone hard.

- **Acorns and hazelnuts** are true nuts.

- **Cola drinks** get their name from the African kola nut, but there are no nuts in them. The flavour is artificial.

- **Some nuts**, such as almonds and walnuts, are not true nuts but the hard stones of drupes (fruit like plums).

- **Brazil nuts** and shelled peanuts are not true nuts but just large seeds.

- **Nuts are** a concentrated, nutritious food – about 50% fat and 10–20% protein. Peanuts contain more food energy than sugar and more protein, minerals and vitamins than liver.

▶ Almonds come from trees native to SW Asia but are now grown all over the world

Photosynthesis

- **Plants use** sunlight to chemically join carbon dioxide gas from the air with water to make sugary food. The process is called photosynthesis.

- **Photosynthesis** occurs in leaves in two special kinds of cell: palisade and spongy cells.

- **Inside the palisade** and spongy cells are tiny packages called chloroplasts. A chloroplast is like a little bag with a double skin or membrane. Each is filled with a jelly-like substance called the stroma in which float various structures, such as lamellae. The jelly contains a chemical called chlorophyll which makes leaves green.

- **The leaf** draws in air containing the gas carbon dioxide through pores called stomata. It also draws water up from the ground through the stem and veins.

- **When the sun** is shining, the chlorophyll soaks up its energy and uses it to split water into hydrogen and oxygen. The hydrogen released from the water combines with the carbon dioxide to make sugar; the oxygen goes out through the stomata.

- **Sugar is transported** around the plant to where it is needed. Some sugar is burned up at once, leaving carbon dioxide and water. This is called respiration.

- **Some sugar is combined** into large molecules called starches, which are easy for the plant to store. The plant breaks these starches down into sugars again whenever they are needed as fuel.

- **Starch** from plants is the main nutrient we get when we eat food such as bread, rice and potatoes. When we eat fruits, cakes or anything else sweet, the sweetness comes from sugar made by photosynthesis.

- **Together** all the world's plants produce about 150 billion tonnes of sugar each year by photosynthesis.

> ★ STAR FACT ★
> The oxygen in the air on which we depend for life was all made by plants during photosynthesis.

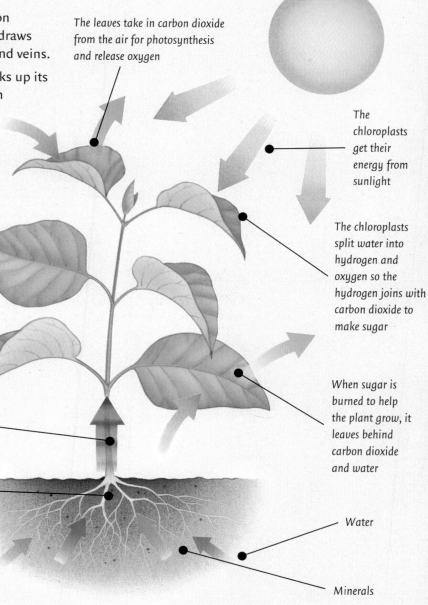

The leaves take in carbon dioxide from the air for photosynthesis and release oxygen

The chloroplasts get their energy from sunlight

The chloroplasts split water into hydrogen and oxygen so the hydrogen joins with carbon dioxide to make sugar

When sugar is burned to help the plant grow, it leaves behind carbon dioxide and water

The minerals are carried up through the plant dissolved in the water

The plant takes up water and minerals from the soil through the roots

Water

Minerals

▶ Every green plant is a remarkable chemical factory, taking in energy from the sun and using it to split water into hydrogen and oxygen. It then combines the hydrogen with carbon dioxide from the air to make sugar, the fuel the plant needs to grow.

Tree facts

▲ *General Sherman in California is the biggest living tree. It is a giant sequoia over 83 m tall and with a trunk 11 m across.*

- **The biggest tree** ever known was the Lindsey Creek Tree, a massive redwood which blew over in 1905. It weighed over 3300 tonnes.
- **The tallest living tree** is the 112 m high Mendocino redwood tree in Montgomery State Reserve, California.

- **The tallest tree** ever known was a Eucalyptus on Watts River, Victoria, Australia, measured at over 150 m in 1872.
- **The great banyan** tree in the Indian Botanical Garden, Calcutta has a canopy covering 1.2 hectares.
- **Banyan trees** grow trunk-like roots from their branches.
- **A European chestnut** known as the Tree of the Hundred Horses on Mt Etna in Sicily had a girth (the distance round the trunk) of 57.9 m in the 1790s.
- **A Moctezuma baldcypress** near Oaxaca in Mexico has a trunk over 12 m across.
- **The world's oldest plant** is the King's Holly in southwestern Tasmania, thought to be 43,000 years old.
- **The ombu tree** of Argentina is the world's toughest tree, able to survive axes, fire, storms and insect attacks.

> ★ **STAR FACT** ★
> The 'Eternal God' redwood tree in Prairie Creek, California is 12,000 years old.

Poisonous plants

◀ *Every single part of the deadly nightshade is highly poisonous and eating a berry will kill you. But in the 1500s ladies would drop extracts in their eyes to make their eyes widen attractively, earning it the name 'belladonna'.*

- **There are thousands** of plants around the world that are poisonous at least in parts.
- **Some parts** of edible plants are poisonous, such as potato leaves and apricot and cherry stones.

- **Some plants** are toxic to eat; some toxic to touch; some create allergic reactions through the air with their pollen.
- **The rosary pea** has pretty red and black seeds often used to make bracelets. But eating one seed can kill a man.
- **Oleanders** are so poisonous that people have been killed by eating meat roasted on an oleander stick.
- **Poison ivy** inflames skin badly if touched.
- **Hemlock** belongs to the parsley family but is highly poisonous. It was said to be the plant used to kill the Ancient Greek philosopher Socrates.
- **Birthwort** is a poisonous vine, but its name comes from its use in the past to help women in childbirth.
- **Crowfoots** such as aconite and hellebore, and spurges such as castor-oil and croton are poisonous.
- **Many useful drugs** are poisons extracted from plants and given in small doses including digitalis from foxgloves, morphine from poppies, atropine from deadly nightshade, quinine, aconite, strychnine and cocaine.

Ferns

- **Ferns** belong to a group of plants called feather plants or pteridophytes, along with club mosses and horsetails.

- **Featherplants** are among the world's most ancient plants, found as fossils in rocks 400 million years old.

- **Coal is made** largely of fossilized featherplants of the Carboniferous Period 360-286 million years ago.

- **There are now** 10,000 species of fern living in damp, shady places around the world.

- **Some ferns** are tiny, with mossy leaves just 1 cm long.

- **Rare tropical tree ferns** can grow up to 25m tall.

- **Fern leaves** are called fronds. When new they are curled up like a shepherd's crook, but they gradually uncurl.

- **Ferns** grow into new plants not from seeds but from spores in two stages.

- **First** spores are made in sacs called sporangia. These are the brown spots on the underside of the fronds. From these spores spread out. Some settle in suitable places.

▶ Most ferns grow on the ground in damp, shady places, but some grow on the leaves or stems of other plants.

- **Second** spores develop into a tiny heart-shaped plant called a prothallus that makes male and female cells. When bathed in rain, the male cells swim to the female cells, fertilizing them. A new root and stem then grow into a proper fern frond and the tiny prothallus dies.

Maize or corn

- **Maize or corn** is the USA's most important crop, and the second most important crop around the world after wheat. Rice is the third.

- **Corn,** like all cereals, is a kind of grass.

- **Corn** was first grown by the Indians of Mexico over 7000 year ago and so came to be called Indian corn by Europeans like Columbus.

- **In the USA** only varieties that give multi-coloured ears are now called Indian corn.

- **The Corn Belt** of the USA grows 40% of the world's corn.

- **American corn** grows up to 3 m tall

- **The ear or head** of a corn plant is called a cob and is covered with tightly packed yellow or white kernels of seeds. The kernels are the part of the plant that is eaten.

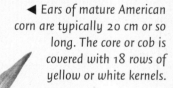

◀ Ears of mature American corn are typically 20 cm or so long. The core or cob is covered with 18 rows of yellow or white kernels.

- **There are seven main kinds** of corn kernel: dent corn, flint corn, flour corn, sweet corn, popcorn, waxy corn and pod corn.

- **Some corn** is ground into flour; some is eaten whole as sweet corn; some is fed to livestock.

- **Popcorn** has no starch, unlike most other corn. When heated, moisture in the kernels turns to steam and expands or pops rapidly.

Cereals

▶ Harvesting wheat and using it to make flour is a surprisingly complex process. The process is still done with simple tools in some parts of the world. But in the developed world, the entire process is largely mechanized.

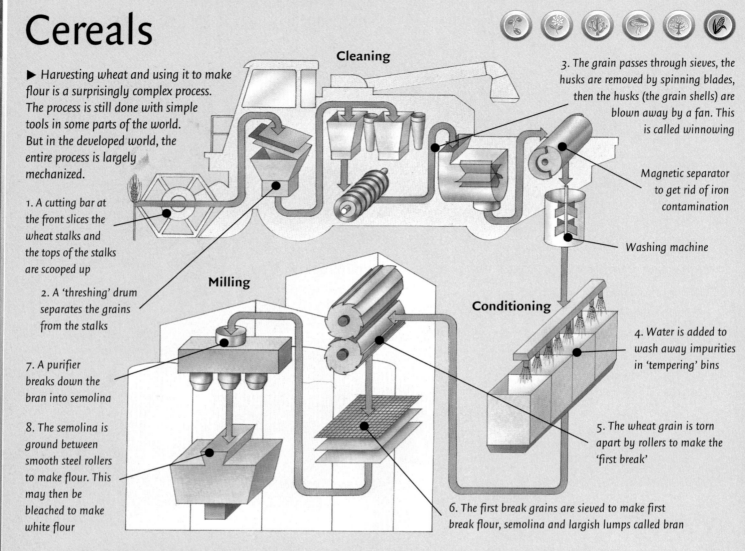

Cleaning

3. The grain passes through sieves, the husks are removed by spinning blades, then the husks (the grain shells) are blown away by a fan. This is called winnowing

Magnetic separator to get rid of iron contamination

Washing machine

1. A cutting bar at the front slices the wheat stalks and the tops of the stalks are scooped up

2. A 'threshing' drum separates the grains from the stalks

Milling

7. A purifier breaks down the bran into semolina

8. The semolina is ground between smooth steel rollers to make flour. This may then be bleached to make white flour

Conditioning

4. Water is added to wash away impurities in 'tempering' bins

5. The wheat grain is torn apart by rollers to make the 'first break'

6. The first break grains are sieved to make first break flour, semolina and largish lumps called bran

- **Cereals** such as wheat, maize, rice, barley, sorghum, oats, rye and millet are the world's major sources of food.

- **Cereals are grasses** and we eat their seeds or grain.

- **The leaves and stalks** are usually left to rot into animal feed called silage.

- **Some grains** such as rice are simply cooked and eaten. Most are milled and processed into foods such as flour, oils and syrups.

- **In the developed world** – that is, places like North America and Europe – wheat is the most important food crop. But for half the world's population, including most people in Southeast Asia and China rice is the staple food.

- **Many grains** are used to make alcoholic drinks such as whisky. A fermentation process turns the starch in the grains to alcohol. Special processing of barley creates a food called malt, which is used by brewers to make beer and lager.

- **Oats** have a higher food value than any other grain.

- **Rye** makes heavy, black bread. The bread is heavy because rye does not contain much gluten which yeast needs to make bread rise.

- **Russia** grows more oats and rye than any other country.

- **Millet** gives tiny seeds and is grown widely in dry regions of Africa and Asia. It was the main crop all over Europe, Asia and Africa in ancient and medieval times.

▶ Wheat flour is used to make everything from pasta to bread.

Cactus

- **Cactuses** are American plants with sharp spines, thick, bulbous green stems and no leaves.

- **Most cactuses** grow in hot, dry regions but a few grow in rainforests and in cold places such as mountain tops.

- **Cactuses** in deserts have a thick, waxy skin to cut water loss to the bare minimum.

◄ *The huge saguaro cactus grows only in the dry foothills and deserts of southern Arizona, southeast California and northwest Mexico.*

★ STAR FACT ★
The stems of the jumping cholla fall off so easily they seem to jump on passers-by.

- **The fat stems** of cactuses hold a lot of water so that they can survive in hot, dry deserts.

- **Because of their moist stems**, cactuses are called succulents.

- **Cactuses have spines** to protect themselves from animals which eat any moist vegetation.

- **Cactuses** have to pollinate just like every flowering plant. So every few years, many produce big colourful blooms to attract insects quickly.

- **Most cactuses** have very long roots to collect water from a large area. The roots grow near the surface to collect as much rainwater as possible.

- **The biggest cactus** is the saguaro, which can grow up to 20 m tall and 1 m thick.

Plankton

- **Plankton** are tiny floating organisms (living things) that are found in both the sea and ponds and lakes.

- **The word 'plankton'** comes from a Greek word meaning 'wandering'.

- **Plankton** is a general term that includes every marine organism too small and weak to swim for itself.

- **The smallest algae** are called plankton, but large floating algae (seaweeds) are not called plankton.

- **Plankton** can be divided into phytoplankton, which are tiny plants, and zooplankton, which are tiny animals, but the division is blurred.

- **Most phytoplankton** are very tiny indeed and so called nannoplankton and microplankton. Zooplankton are generally bigger and called macroplankton.

- **Green algae** that give many ponds a bright green floating carpet are plankton.

- **Phytoplankton** get their energy by photosynthesis just like other plants.

- **Countless puffs** of oxygen given out by plankton early in Earth's history gave the air its vital oxygen.

- **Plankton** is the basic food of all large ocean animals.

▼ *Diatoms are at the beginning of the ocean food chain. They use the Sun's energy for growth.*

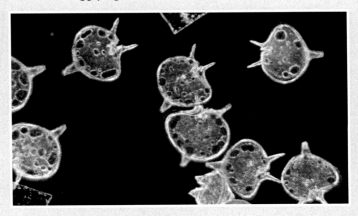

Garden flowers

▲ Most gardens now have a mix of trees and shrubs, mixed beds of herbaceous flowers and early-flowering bulbs such as crocuses.

- **All garden flowers** are descended from plants that were once wild, but they have been bred over the centuries to produce flowers quite unlike their wild relatives.

- **Garden flowers** like tea roses, created by cross-breeding two different species, are called hybrids.

- **Garden flowers** tend to have much bigger blooms and last for longer than their wild cousins.

- **By hybridization** gardeners have created colours impossible naturally, such as black roses.

- **Ornamentals** are flowers cultivated just for show.

- **Gardeners** try to mix flowers that bloom at different times so that the garden is always full of colour.

- **18th century botanist** Carl Linnaeus made a clock by planting flowers that bloomed at different times of day.

- **The earliest flowerbeds** were the borders of flower tufts Ancient Persians grew along pathways.

- **A herbaceous border** is a traditional flowerbed planted with herbaceous perennial flowers like delphiniums and chrysanthemums. It flowers year after year.

> ★ STAR FACT ★
> Herbaceous borders were invented by Kew gardener George Nicolson in the 1890s.

Cocoa

- **Cocoa beans** are the fruit of the cacao tree.

- **Cocoa beans** are called cocoa beans and not cacao beans because of a spelling mistake made by English importers in the 18th century when chocolate first became popular.

- **Cocoa beans** are the seeds inside melon-shaped pods about 30 cm long.

- **Cacao trees** came originally from Central America. Now they are grown in the West Indies and West Africa too.

- **Chocolate** is made by grinding the kernels of cocoa beans to a paste called chocolate liquor. The liquor is hardened in moulds to make chocolate.

- **Cooking chocolate** is bitter. Eating chocolate has sugar and, often, milk added.

◀ The cacao tree is a tall tropical tree growing up to about 8 m. The seeds used to make cocoa are small beans inside the melon-sized pod.

- **Cocoa powder** is made by squeezing the cocoa butter (fat) from chocolate liquor then pulverizing it.

- **When Spanish explorer** Hernán Cortés reached the court of Moctezuma (Aztec ruler of Mexico in 1519) he was served a bitter drink called *xocoatl*. The people of Central America had regarded *xocoatl* as a sacred drink since the time of the Mayans.

- **In the 1600s** Europeans began to open fashionable chocolate houses to serve *xocoatl* as hot chocolate sweetened with sugar. In the 1700s, the English began adding milk to improve the flavour.

- **'Cacao'** is a Mayan word for 'bitter juice'; chocolate comes from the Mayan for 'sour water'.

Plants and water

- **Plants cannot survive** without water. If they are deprived of water, most plants will wilt and die very quickly – although some desert plants manage to get by on very little indeed.

- **Nearly all plants** are almost 70% water, and some algae are 98% water.

- **In plants** water fills up the tiny cells from which they are made, and keeps them rigid in the same way as air in a balloon.

- **For a plant** water also serves the same function as blood in the human body. It carries dissolved gases, minerals and nutrients to where they are needed.

▲ *Plants need regular watering to keep them fresh and healthy.*

- **Some water** oozes from cell to cell through the cell walls in a process called osmosis.

- **Some water** is piped through tubes called xylem. These are the fine veins you can often see on leaves.

- **Water in xylem** is called sap and contains many dissolved substances besides water.

- **Plants lose water** by transpiration. This is evaporation through the leaf pores or stomata.

- **As water is lost** through the stomata, water is drawn up to replace it through the xylem.

- **If there is too little** water coming from the roots, the cells collapse and the plant wilts.

Algae

- **Algae** are simple organisms that live in oceans, lakes, rivers and damp mud.

- **Some algae** live inside small transparent animals.

- **Algae vary** from single-celled microscopic organisms to huge fronds of seaweed (brown algae) over 60 m long.

- **The smallest** float freely, but others, such as seaweeds, need a place to grow like a plant.

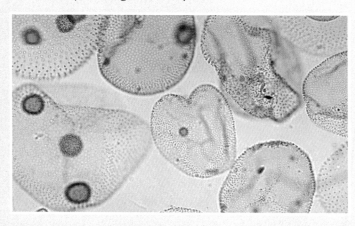

- **Algae** are so varied and often live very differently from plants, so scientists put them not in the plant kingdom but in a separate kingdom called the *Proctista*, along with slime moulds.

- **The most ancient** algae are called blue-green algae or cyanobacteria and are put in the same kingdom as bacteria. They appeared on Earth 3 billion years ago.

- **Algae** may be tiny but they are a vital food source for creatures from shrimps to whales, and they provide most of the oxygen water creatures need for life.

- **Green algae** are found mostly in freshwater. The green is the chlorophyll that enables plants to get their energy from sunlight.

- **Green algae** called *Spirogyra* form long threads.

- **Red or brown algae** are found in warm seas. Their chlorophyll is masked by other pigments.

◄ *Volvox are green algae that live in colonies about the size of a pinhead, containing up to 60,000 cells.*

Grapes

- **Grapes** are juicy, smooth-skinned berries that grow in tight clusters on woody plants called vines.

- **Grapes** can be black, blue, green, purple, golden or white, depending on the kind.

- **Some grapes** are eaten fresh and some dried as raisins, but 80% are crushed to make wine.

- **Grapes** are grown all round the world in places where there are warm summers and mild winters, especially in France, Italy, Spain, Australia, Chile, Romania, Georgia, South Africa and California.

- **Among the best wine grapes** are the Cabernet Sauvignon and Chardonnay for white wine and the Pinot Noir for red wine.

◄ *Grapes have been cultivated since the earliest times. Purple grapes like these will be used to make red wine.*

- **The Ancient Egyptians** made wine from grapes 5000 years ago.

- **Grapes are made** into wine by a process called fermentation.

- **Grapes** for eating fresh are called table grapes and are bigger and sweeter than wine grapes. Varieties include Emperor, red Tokay, green Perlette and black Ribier.

- **Grapes grown** for raisins are seedless. The best known is Thompson's seedless, sometimes called the sultana.

- **Grapevines** are grown from cuttings. They start to give fruit after three or four years and may bear fruit for a century. Each vine usually gives 10-35 kg of grapes.

Growing from seed

     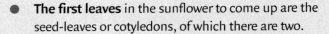

► *This illustration shows some of the stages of germination, as a plant grows from a seed – here a bean seed.*

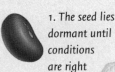

1. The seed lies dormant until conditions are right

- **When seeds mature**, they contain the germ (embryo) of a new plant and the food needed to grow it.

- **The seed lies dormant** (inactive) until conditions are right for it to germinate (grow into a plant) – perhaps when it begins to warm up in spring.

- **Poppy seeds** can lie buried in soil for years until brought to the surface by ploughing, allowing them to grow.

- **Scientists once grew** plants from lotus seeds that were 10,000 years old.

- **A seed needs** water and warmth to germinate.

- **When a seed germinates** a root (or radicle) grows down from it and a green shoot (or plumule) grows up.

2. The seed sends a root down and a shoot up

- **The first leaves** in the sunflower to come up are the seed-leaves or cotyledons, of which there are two.

- **Only certain parts** of a plant, called meristems, can grow. These are usually the tips of shoots and roots.

- **Because** a plant grows at the tips, shoots and roots mainly get longer rather than fatter. This is called primary growth.

- **Later in life** a plant may grow thicker or branch out.

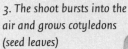

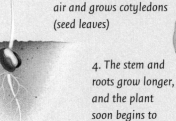

3. The shoot bursts into the air and grows cotyledons (seed leaves)

4. The stem and roots grow longer, and the plant soon begins to grow new leaves

Prairie and steppe

- **Grasslands in cool parts** of the world are called prairies or steppes. There is not enough rain all year round for trees to grow.

- **Prairies** are the grasslands of North America. Steppes are the grasslands of Russia. Every region has its own name for grasslands, such as the veld in South Africa and pampas in South America. But now grasslands anywhere with tall grass are usually called prairies and grasslands with shorter grass are usually called steppes.

- **Hundreds of kinds** of grass grow in prairies. In moist areas in North America, there are grasses like switch grass, wild rye, Indian grass and big bluestem. In drier areas, the main grasses are dropseeds, little bluestem, June grass, needlegrass and blue grama. Slough grass grows in marshland. The state of Kentucky is famous for its bluegrass.

▼ *When European pioneers first saw the American prairies in the 19th century, they described them as 'a sea of grass, stretching to the horizon'. Now, corn and wheat fields and cattle ranches cover most of them. Wild prairies like this are now very rare.*

- **Meadow grass** is the most common of all grasses, found on grasslands all over the world – and in garden lawns.

- **Shrubs** such as prairie roses often grow amid the grass, while oaks, cottonwoods and willows grow near rivers.

- **The many prairie flowers** include blazing stars, coneflowers, sunflowers, asters and goldenrods.

- **Eurasian grasslands** bloom with vetches, trefoils, worts, orchids and many herbs.

- **Grasslands cover** nearly a quarter of the Earth's land surface.

- **When grasslands** are destroyed by farming, the soil can be blown away by the wind as in the dust bowl of N. America in the 1900s.

> ★ **STAR FACT** ★
> Prairies and steppes typically have very dark soils such as chernozems. The word *chernozem* is Russian for 'black earth'.

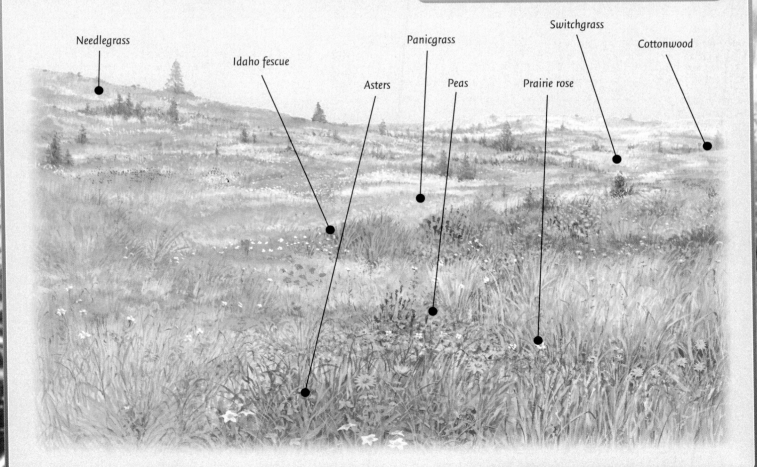

Needlegrass

Idaho fescue

Asters

Panicgrass

Peas

Prairie rose

Switchgrass

Cottonwood

Palm trees

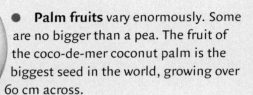

▶ Date palms produce several clusters of 600-1700 dates towards the end of the year, each year, for about 60 years.

Palm tree

Date

Raffia Palm

- **Palms** are a group of 2780 species of tropical trees and shrubs.

- **Palms** have a few very large leaves called fronds.

- **The fronds** grow from the main bud at the top of a tall thin trunk.

- **If the main bud** at the top of the trunk is damaged, the tree will stop growing and die.

- **Palm trunks** do not get thicker like other trees; they simply grow taller.

- **Some palms** have trunks no bigger than a pencil; others are 60 m high and 1 m across.

- **Palm fruits** vary enormously. Some are no bigger than a pea. The fruit of the coco-de-mer coconut palm is the biggest seed in the world, growing over 60 cm across.

- **Palm trees** are a very ancient group of plants, and fossil palms have been found dating back 100 million years to the time of the dinosaurs.

- **Date palms** have been cultivated in the hottest parts of North Africa and the Middle East for at least 5000 years. Muslims regard it as the tree of life.

> ★ STAR FACT ★
> The world's largest leaves are those of the Raffia palm, which grow up to 20 m long.

Coastal plants

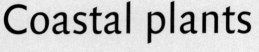

- **Plants** that grow on coasts must be able to cope with exposure to wind and salt spray, and thin, salty soils.

- **Plants** that can tolerate salt are called halophytes.

- **Spray halophytes** can tolerate occasional splashing.

- **True halophytes** can tolerate regular immersion when the tide comes in.

◀ Sea pinks are also known as thrift because they 'thrive' all the year round on the most exposed cliffs.

- **The annual seablite** is a true halophyte that lives in between the tides. The word 'blite' comes from an old English word for spinach.

- **The rock samphire's** name comes from St Pierre (St Peter) who was known as the rock. The plant clings to bare rock faces. Samphire was once a popular vegetable and poor people risked their lives to collect it from cliffs.

- **The droppings** of sea birds can fertilize the soil and produce dense growths of algae and weeds such as dock.

- **Lichens** on rock coasts grow in three colour bands in each tidal zone, depending on their exposure to salt.

- **Grey 'sea ivory' lichen** grows above the tide; orange lichens survive constantly being splashed by waves; black lichens grow down to the low water mark.

- **On pebble and shingle beaches** salt-tolerant plants like sea holly, sea kale and sea campion grow.

Lilies

▲ *Lilies are one of the most popular garden flowers and have been cultivated in a wide range of colours.*

- **Lilies** are one of the largest and most important flower families, containing about 4000 species.

- **Lilies** are monocots (which means a single leaf grows from their seeds) and give their name to the entire group of monocots – liliopsidae.

- **The lily family** includes many flowers called lilies but also asparagus and aloes.

- **Hyacinths** belong to the lily family.

- **Lilies** grow from bulbs to produce clusters of bright trumpet-shaped flowers on tall stems. Each flower has six petals.

- **Lily-of-the-valley** has tiny white bell-shaped blooms. According to superstition, anyone who plants it will die within a year.

- **Lilies-of-the-valley** are famous for their fragrance. They are used to scent soaps and perfumes.

- **Easter lilies** are large trumpet-shaped white lilies that have come to symbolize Easter.

- **Leopard lilies** grow in the western coastal states of the United States. They have red-orange flowers spotted with purple.

- **The Madonna lily** is a lily planted in August that lives throughout the winter.

Cotton

- **Cotton** is a fibre that comes from the cotton plant.

- **The cotton plant** is a small shrub that grows in tropical and subtropical climates.

- **Cotton plants** are annuals and are planted fresh each spring.

- **Cotton plants** grow seed pods called bolls, containing 20–40 seeds – each covered with soft, downy hairs or fibres.

- **As bolls ripen** they burst open to reveal the mass of fluffy fibres inside.

- **When separated** from the seeds, the fluff is known as cotton lint.

- **Cotton seeds** are processed to make oil, cattle cake and fertilizer.

- **There are 39 species** of cotton plant, but only four are cultivated: the upland, Pima, tree and Levant.

- **Upland plants** give 90% of the world's cotton.

▲▶ *The bolls picked for cotton develop from the seed pod left when the petals of the cotton flower drop off in summer.*

- **Upland** and Pima both came from the Americas, unlike tree and Levant, which are from the Middle East and Africa.

Eucalyptus trees

- **Eucalyptus trees** make up a group of over 400 species of Australian trees. They grow fast and straight, and often reach tremendous heights.

- **Eucalyptus trees** grow best in warm places with marked wet and dry seasons.

- **In winter** eucalyptus trees simply stop growing and produce no new buds.

- **Eucalyptus trees** in California were grown originally from seeds that came from Tasmania.

- **Australians** often call eucalyptus trees gum trees or just gums.

- **Eucalyptus leaves** give eucalyptus oil, used as vapour rubs for people with colds.

◄ *Eucalyptus trees have long, narrow, leathery leaves which are cut, pressed and then steamed to make eucalyptus oil.*

- **The most important** tree grown for oil is the Blue mallee or blue gum. Blue gum trees are the most widespread in North America.

- **Some eucalyptus trees** give Botany Bay kino, a resin used to protect ships against worms and other animals that make holes in their hulls.

- **The jarrah** is an Australian eucalyptus that gives a red wood rather like mahogany. Other eucalyptus woods are used to make everything from boats to telegraph poles.

> ★ STAR FACT ★
> Eucalyptus trees can grow to over 90 m tall – taller than any trees but Californian redwoods.

Green vegetables

- **Green vegetables** are the edible green parts of plants, including the leaves of plants such as cabbages and the soft stems of plants like asparagus.

- **Cabbages** are a large group of green vegetables called the brassicas.

- **Cabbages were** originally developed from the sea cabbage (*Brassica oleracea*) which grew wild near sea coasts around Europe.

- **Kale and collard** are types of cabbage with loose, open leaves.

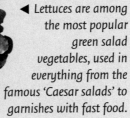

◄ *Lettuces are among the most popular green salad vegetables, used in everything from the famous 'Caesar salads' to garnishes with fast food.*

- **Common and savoy** cabbages are cabbages with leaves folded into a tight ball. Brussel sprouts are cabbages with lots of compact heads.

- **Cauliflower and broccoli** are cabbages with thick flowers. Kohlrabi is a cabbage with a bulbous stem.

- **The leaves of green vegetables** are rich in many essential vitamins including vitamin A, vitamin E and folic acid (one of the B vitamins).

- **Spinach** looks a little like kale, but it is actually a member of the goosefoot family, rich in vitamins A and C, and also in iron. The discovery of the iron content made spinach into the superfood of the cartoon hero Popeye in the mid 20th century.

- **Asparagus** belongs to the lily family. Garden asparagus has been prized since Roman times.

- **In Argenteuil** in France, asparagus is grown underground to keep it white. White asparagus is especially tender and has the best flavour.

Marshes and wetlands

- **There are two kinds of marsh:** freshwater marshes and saltwater marshes.

- **Freshwater marshes** occur in low-lying ground alongside rivers and lakes where the water level is always near the soil surface.

- **Freshwater marshes** are dominated by rushes, reeds and sedges.

- **Sedges** are like grass but have solid triangular stems. They grow in damp places near the water's edge.

- **Rushes** have long cylindrical leaves and grow in tussocks in damp places along the bank.

- **Reeds** are tall grasses with round stems, flat leaves and purplish flowers. They grow in dense beds in open water.

- **Free-floating** plants like duckweed and frogbit are common in marshes. In rivers they'd be washed away.

- **Water horsetails** are relics of plants that dominated the vast swamps of the Carboniferous Period some 300 million years ago.

▲ Reeds and floating duckweed thrive in open water in marshes.

- **Saltwater marshes** are flooded twice daily by salty seawater. Cordgrasses and salt-meadow grass are common. Reeds and rushes grow where it is least salty.

- **Where mud is firm,** glasswort and seablite take root. Further from the water sea aster and purslane grow. On high banks, sea lavender, sea plantain and thrift bloom.

Bulbs and suckers

- **Annuals and biennials** only grow once, from a seed. Many perennials die back and grow again and again from parts of the root or stem. This is called vegetative propagation.

- **Plants such as lupins** grow on the base of an old stem. As the plant ages, the stem widens and the centre dies, leaving a ring of separate plants around the outside.

    - **Plants such as irises** sprout from thick stems called rhizomes. These grow sideways beneath the ground.

        - **If the end** of a rhizome swells up it forms a lump called a tuber.

            - **Potatoes** are the tubers of the potato plant.

Bulb

Corm

Tuber

Rhizome

- **Flowers like crocuses** and gladioli have a bulbous base to their stem. This is called a corm.

- **Bulbs like those** of tulips, daffodils and onions look like corms, but they are actually made of leaf parts rather than the stem. This is why they have layers.

- **Garlic bulbs** are separated into four or five segments called cloves.

- **In winter,** rhizomes, tubers, corms and bulbs act as food stores. In spring they provide the energy to grow new leaves.

- **Plants can also** propagate (grow new plants) by sending out long stems that creep over the ground called runners or under the ground (suckers).

Seed dispersal

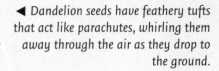

◀ *Dandelion seeds have feathery tufts that act like parachutes, whirling them away through the air as they drop to the ground.*

- **After maturing** seeds go into a period called dormancy. While they are in this state they are scattered and dispersed.

- **Some scattered seeds** fall on barren ground and never grow into plants. Only those that fall in suitable places will begin to grow.

- **Some seeds** are light enough to be blown by the wind. The feathery seed cases of some grasses are so light they can be blown several kilometres.

- **Many seeds and fruits** have wings to help them whirl through the air. Maple fruits have wings. So too do the seeds of ashes, elms and sycamores.

- **Seeds** like dandelions, cottonwoods and willows have fluffy coverings, so they drift easily on the wind.

- **Some seeds** are carried by water. Coconut seeds can float on the sea for thousands of kilometres.

- **Many fruits and seeds** are dispersed by animals.

- **Some fruits** are eaten by birds and other animals. The seeds are not digested but passed out in the animal's body waste.

- **Some seeds** stick to animal fur. They have burrs or tiny barbs that hook on to the fur, or even a sticky coating.

- **Some fruits,** like geraniums and lupins, simply explode, showering seeds in all directions.

▶ *Sycamore seeds have wings to help them spin away on the wind.*

Mountain plants

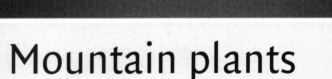

- **Conditions get colder,** windier and wetter higher up mountains, so plants get smaller and hardier.

- **On lower slopes** conifers such as pines, firs, spruces and larches often grow.

- **Above a certain height,** called the tree-line, it gets too cold for trees to grow.

- **In Australia,** eucalyptus trees grow near the tree-line. In New Zealand, Chile and Argentina southern beeches grow.

> ★ STAR FACT ★
> On Mt Kenya in Africa, huge dandelion-like plants called giant groundsels grow as big as trees.

- **Above the tree-line** stunted shrubs, grasses and tiny flowers grow. This is called alpine vegetation.

- **Alpine flowers** like purple and starry saxifrage have tough roots that grow into crevices and split the rocks.

- **There are few insects** high up, so flowers like saxifrage and snow gentian have big blooms to attract them.

- **To make the most** of the short summers, the alpine snowbell grows its flower buds the previous summer, then lets the bud lie dormant through winter under snow.

- **Alpine flowers** such as edelweiss have woolly hairs to keep out the cold. Tasmanian daisies grow in dense cushion-shapes to keep warm.

◀ *As you go higher up a mountain, the trees of the lower slopes thin out. At the top, only mosses and lichens grow.*

Tropical grassland

- **Tropical grasslands** are regions in the tropics where there is not enough rain half the year for trees to grow.

- **Grasses** in tropical grasslands tend to grow taller and faster than grasses in cooler regions.

- **Grass stalks** may be eaten by grazing animals, burned by bush fires or dry out, but roots survive underground.

- **In Africa** grasses include 3 m-tall elephant grasses. In Australia, they include tall spear grass and shorter kangaroo grass. In South America, there are plants called bunch grasses and species such as Briza.

- **Most tropical grasslands** are scattered with bushes, shrubs and trees. In Africa, typical trees include hardy broad-leaved trees such as curatella and byrsonima.

- **Many grassland trees** are said to be sclerophyllous. This means they have tough leaves and stems to save water.

- **In drier regions** acacias and other thorn trees are armed with spines to protect them against plant-eating animals. The thorns can be up to 50 cm long.

▲ In East Africa, the grassland is called savanna, and this name is often used for tropical grassland everywhere.

- **In damper places** palm trees often take the place of the thorn trees.

- **Baobab trees** are East African trees with massive trunks up to 9 m across which act as water stores.

- **Baobab trees** look so odd that Arab legend says the devil turned them upside down so their roots stuck up in the air.

Conifers

- **Conifers** are trees with needle-like, typically evergreen leaves that make their seeds not in flowers but in cones.

- **With gingkos and cycads** they make up the group of plants called gymnosperms, all of which make their seeds in cones.

- **The world's tallest tree,** the redwood, is a conifer.

- **The world's most massive tree,** the giant sequoia, is a conifer.

- **One of the world's oldest trees** is the bristlecone pine of California and Nevada, almost 5000 years old.

- **The world's smallest trees** are probably conifers including natural bonsai cypresses and shore pines which reach barely 20 cm when fully grown.

- **Many conifers** are cone-shaped, which helps them shed snow in winter.

- **The needle-like shape** and waxy coating of the leaves helps to save water.

- **The needles of some pines** can grow up to 30 cm long. But the biggest needles ever were those of the extinct *Cordaites*, over 1 m long and 15 cm wide.

- **Conifers** grow over most of the world, but the biggest conifer forests are in places with cold winters, such as north Siberia, northern North America and on mountain slopes almost everywhere.

◄ Most conifers are instantly recognizable from their conical shapes, their evergreen, needle-like leaves and their dark brown cones.

INDEX

Acknowledgements

Artists: Lisa Anderson (Advocate), Roger Goode (Beehive illustration), Roger Kent (Illustration Ltd.), Rob Jakeway, Janos Marffy, Terry Riley, Martin Sanders, Peter Sarson, Rudi Vizi, Mike White (Temple Rogers), John Woodcock.

The publishers would like to thank the following sources for the photographs used in this book:

Page 27 (C/R) Science Photo Library; Page 31 (C) Corbis; Page 35 (T/R) The Stock Market; Page 39 (T/L) Science Photo Library; Page 50 (B/L) The Stock Market; Page 58 (T/R) Science Photo Library; Page 75 (T/R) CORBIS; Page 77 (T/L) Steve Lindridge, Eye Ubiquitous/CORBIS; Page 80 (B/L) Richard T. Nowitz/CORBIS; Page 91 (B/L) CORBIS; Page 104 (B/L) Jim Sugar Photography/CORBIS; Page 107 (B/R) Joseph Sohm, ChromoSohm Inc./CORBIS; Page 109 (T/R) In-press photography; Page 115 (T/R) Wolfgang Kaehler/CORBIS Page 132 (T/L) Wolfgang Kaehler/CORBIS; Page 132 (B/L) Patrick Johns/CORBIS; Page 135 (B/L) Wolfgang Kaehler/CORBIS; Page 138 (B/C) Paul A. Souders/CORBIS; Page 140 (T/C) Guy Stubbs, Gallo Images/CORBIS; Page 145 (B/R) Panos Pictures; Page 150 (B/L) Catherine Karnow/CORBIS; Page 157 (T/R) Ted Atkinson, Eye Ubiquitous/CORBIS; Page 163 (B/L) John Batholomew/CORBIS; Page 164 (B/L) Panos Pictures; Page 165 (B/R) Guy Stubbs, Gallo Images/CORBIS; Page 168 (T/L) David Cumming, Eye Ubiquitous/CORBIS; Page 171 (B/C) Jeffrey L. Rotman/CORBIS; Page 173 (C/L) Daewoo Cars Ltd.; Page 174 (B/L) Owen Franken/CORBIS; Page 175 (T/R) London Aerial Photo Library/CORBIS; Page 177 (B/R) Fancoise de Mulder/CORBIS; Page 178 (T/L) Kelly-Mooney Photography/CORBIS; Page 180 (B/L) Charles & Josette Lenars/CORBIS. Page 192 (T/L) Jo Brewer; Page 196 (T/R) Gunter Marx/CORBIS; Page 201 (T/R) Bob Gibbons/Holt Studios; Page 201 (B/L) Jo Brewer; Page 202 (T/L) Raymond Gehman/CORBIS; Page 204 (T/C) Scott T. Smith/CORBIS; Page 206 (T/R) Woflgang Kaehler/CORBIS; Page 207 (T/R) Jo Brewer; Page 210 (B/C) John Holmes, Frank Lane Picture Agency/CORBIS; Page 211 (B/L) Jo Brewer; Page 214 (T/L) Sally A. Morgan, Ecoscene/CORBIS; Page 217 (T/L) Richard T. Nowitz/CORBIS; Page 220 (T/R) Jo Brewer; Page 220 (B/L) Steve Austin, Papilio/CORBIS; Page 228 (T/R) Wayne Lawler, Ecoscene/CORBIS; Page 232 (B/L) Sally A. Morgan, Ecoscene/CORBIS; Page 233 (B/R) Douglas P. Wilson, Frank Lane Picture Agency/CORBIS; Page 239 (B/R) Roger Wood/CORBIS.

All other photographs are from MKP Archives